International Conference on New Energy and Power Grid (NEPG 2025)

Journal of Physics: Conference Series Volume 3031

Nanjing, China
21-23 March 2025

ISBN: 979-8-3313-2266-3
ISSN: 1742-6588

Printed from e-media with permission by:

Curran Associates, Inc.
57 Morehouse Lane
Red Hook, NY 12571

Some format issues inherent in the e-media version may also appear in this print version.

Institute of Physics
Dirac House, Temple Back
Bristol BS1 6BE UK

Phone: 44 1 17 929 7481
Fax: 44 1 17 920 0979

techtracking@iop.org

Additional copies of this publication are available from:

Curran Associates, Inc.
57 Morehouse Lane
Red Hook, NY 12571 USA
Phone: 845-758-0400
Fax: 845-758-2633
Email: curran@proceedings.com
Web: www.proceedings.com

TABLE OF CONTENTS

Journal of Physics: Conference Series **3031** (2025) 011001 doi:10.1088/1742-6596/3031/1/011001

Preface

With the accelerating transition toward low-carbon economies and digital power systems, research on renewable energy, intelligent grids, and sustainable energy management has entered a new era of interdisciplinary innovation. Against this backdrop, the 2025 International Conference on New Energy and Power Grid (NEPG 2025) was successfully held online on March 21, 2025, bringing together over 60 participants from academia, industry, and research institutions worldwide. The conference served as a dynamic platform for sharing insights, forging collaborations, and advancing the global energy transition.

This year's conference received 58 high-quality submissions covering a diverse range of topics, including but not limited to new energy systems, smart grids, energy storage, power electronics, grid integration, and sustainable development. Following a rigorous peer review process, 21 papers were accepted for publication in these proceedings, each selected for their scientific contribution and relevance to the evolving energy landscape.

NEPG 2025 was honored to feature four distinguished keynote speakers whose expertise enriched the conference with forward-looking perspectives. Prof. Shunli Wang (Smart Energy Storage Institute, China) shared insights into secure and scalable energy storage technologies; Assoc. Prof. Jiashen Teh (Universiti Sains Malaysia) discussed the integration of renewable energy into modern grids; Prof. Yacine Amara and Dr. Amina Bensa (Université Le Havre Normandie, France) presented cutting-edge research on advanced electric machines and multiphysics modeling for sustainable energy applications. Their contributions sparked meaningful dialogue, inspired interdisciplinary collaboration, and set a high academic tone for the event.

We extend our sincere appreciation to all the authors, reviewers, speakers, and attendees for their valuable contributions and enthusiastic participation. Our gratitude also goes to the organizing committee, whose professionalism and commitment ensured the smooth running and high standard of the event. We are confident that the knowledge shared and relationships formed at NEPG 2025 will drive meaningful advances in new energy and power grid technologies.

We look forward to welcoming you to the next edition of the conference as we continue to build a cleaner, smarter energy future together.

NEPG 2025 Conference Committee

Journal of Physics: Conference Series **3031** (2025) 011001 doi:10.1088/1742-6596/3031/1/011001

Committee Member

Conference Chair
Prof. Shunli Wang, Smart Energy Storage Institute, China

Organizing Committee Chair
Assoc. Prof. Jiashen The, Universiti Sains Malaysia, Malaysia

Technical Program Committee Chair
Prof. dr. eng. Gheorghe Grigoraș, "Gheorghe Asachi" Technical University of IAŞI, Romania

Publication Chair
Prof. Farhad Shahnia, Murdoch University, Australia
Assoc. Prof. Muhyiddine Jradi, The Maersk Mc-Kinney Moller Institute SDU Center for Energy Informatics, Denmark

Technical Program Committee Members
Researcher. Tianbiao He, Zhejiang University, China
Prof. Mohd Junaidi Abdul Aziz, Universiti Teknologi Malaysia, Malaysia
Prof. Maciej Dzikuć, University of Zielona Góra, Poland
Prof. Jahangir Hossain, University of Technology Sydney, Australia
Senior Engineer Xiaonan Yang, China Electric Power Research Institute, China
Assoc. Prof. Jingsong Li, Dalian University of Technology, China

Organizing Committee Members
Prof. MÜSLÜM ARICI, Kocaeli University, Turkey
Prof. Humayun Kabir, University of DhaKA, India
Assoc. Prof. Manoj Khandelwal, Federation University Australia, Australia
Assoc. Prof. Kwan Yiew Lau, Universiti Teknologi Malaysia, Malaysia
Assoc. Prof. Bhargav Appasani, KIIT School of Electronics Engineering, India
Dr. Yuefeng Liao, Zhengzhou University, China

Journal of Physics: Conference Series **3031** (2025) 011002 doi:10.1088/1742-6596/3031/1/011002

Peer Review Statement

All papers published in this volume have been reviewed through processes administered by the Editors. Reviews were conducted by expert referees to the professional and scientific standards expected of a proceedings journal published by IOP Publishing.

- **Type of peer review:** Single Anonymous
- **Conference submission management system:** Morressier
- **Number of submissions received:** 58
- **Number of submissions sent for review:** 43
- **Number of submissions accepted:** 21
- **Acceptance Rate (Submissions Accepted / Submissions Received × 100):** 36.2
- **Average number of reviews per paper:** 2
- **Total number of reviewers involved:** 6
- **Contact person for queries:**
  Name: Xuexia Ye
  Email: xx.ye@keoaeic.org
  Affiliation: AEIC Academic Exchange Information Centre

Journal of Physics: Conference Series **3031** (2025) 012001 doi:10.1088/1742-6596/3031/1/012001

Day-ahead optimal scheduling method for hydrogen-electric coupling systems considering source-load uncertainty

Cheng Zhang [1], Wei Fang [1,*], Banghua Du [2], Qicheng Zhang [1], Xiaolan Dai [1], Hui Wu [1]

[1] School of Automation, Wuhan University of Technology, Wuhan 430070, China
[2] Hubei Key Laboratory of Advanced Technology for Automotive Components, Wuhan University of Technology, Wuhan 430070, China

[*]Corresponding author's e-mail: fangwei49@whut.edu.cn

Abstract. In recent years, the concept of microgrids has gained significant traction. However, uncertainties in source-load dynamics within microgrid systems have introduced critical challenges such as scheduling complexities and cumulative operational errors. Furthermore, the increasing integration of new energy vehicles imposes substantial impacts on microgrid stability under high-load conditions, underscoring the imperative for accurate energy prediction and dispatch. To address these challenges, this study proposes a hybrid deep learning-based day-ahead scheduling optimization system for hydrogen-electricity coupled microgrids. First, mathematical models of core internal components are established. On the generation side, a fused TCN with an attention mechanism and GRU is developed to predict renewable energy outputs and pre-optimize scheduling plans. On the demand side, Monte Carlo simulations combined with K-means clustering are employed to classify and forecast EV charging patterns based on historical behavioral data. Finally, an optimization model is formulated with total operational cost minimization as the objective function to generate day-ahead scheduling strategies. Extensive simulation results demonstrate that the proposed hybrid deep learning framework achieves a 34.8% reduction in MAE compared to traditional deep learning benchmarks while reducing scheduling costs by 5% relative to non-prediction-based approaches.

1. Introduction

With the growing global demand for clean energy, the concept of microgrids has gained significant popularity as independent systems that integrate DERs — such as wind, solar, and fuel cells — with loads and energy storage devices. These systems can operate in both grid-connected and islanded modes [1]. Notably, solar and wind power generation has experienced remarkable growth in recent years, emerging as the fastest-growing contributors to electricity generation since the beginning of the 21st century.

In recent years, numerous researchers have explored the feasibility of wind and solar photovoltaic power generation for hydrogen production and energy supply. For instance, Li et al. [2] investigated a game theory-based power coordination management method for a hybrid hydrogen-battery energy storage system (HBHS) to achieve zero-net-emission targets in building systems. Ceylan and Devrim [3] proposed a hydrogen-based hybrid energy system that utilizes solar and wind power generation to provide electricity in both off-grid and grid-connected modes. Marocco et al. [4] introduced an off-grid community power system design optimized by the particle swarm optimization (PSO) algorithm for component sizing, resulting in an economically feasible and sustainable hydrogen storage system design. Additionally, a significant body of literature has examined the uncertainties associated with EVs and their impacts on power systems. For example, Zhang et al. [5] employed an optimized K-means clustering method to aggregate EV fleets with similar characteristics, ensuring the stability and flexibility of microgrid systems. However, critical research gaps persist in microgrid scheduling optimization, particularly concerning the prediction accuracy degradation caused by supply-side renewable intermittency and the demand-side uncertainties stemming from stochastic user behaviors. Current approaches face dual limitations: (1) Model-based generation forecasts neglect weather-related nonlinearities; (2) EV charging predictions struggle with user behavior uncertainties, causing variable load fluctuations. Furthermore, few studies holistically assess the bidirectional interdependencies between supply-side prediction errors and demand-side EV load volatility in co-optimizing microgrid dispatch economics.

This paper proposes a hydrogen electricity hybrid microgrid day-ahead scheduling plan that considers multiple uncertainties on the source and load side. First, the TCN-GRU deep learning method integrating the attention mechanism is used to predict the day-ahead wind, photovoltaic, and load values. Then, the probability density function integrating the behavior characteristics of new energy owners is used to conduct the Monte Carlo simulation for scene generation. Finally, the K-means method is used to classify and process the features of the scene generation, and the quantitative microgrid cost is used as the index for day-ahead optimal scheduling.

2. System configuration and mathematical modeling

2.1. System configuration

This study's system configuration derives from an integrated hydrogen energy demonstration project. As shown in Figure 1, the hybrid power network primarily utilizes photovoltaic generation, supplemented by wind and hydrogen power. Designed to meet electricity demands through clean energy utilization, it covers the full energy lifecycle from production to consumption. Key components include PV arrays, wind turbines, PEMEL, PEMFC, hydrogen storage tanks, and batteries. Grid connectivity ensures reliability during low renewable generation. The system converts renewable energy through wind/PV infrastructure, stores surplus power, and enables bidirectional grid exchange to optimize operational efficiency.

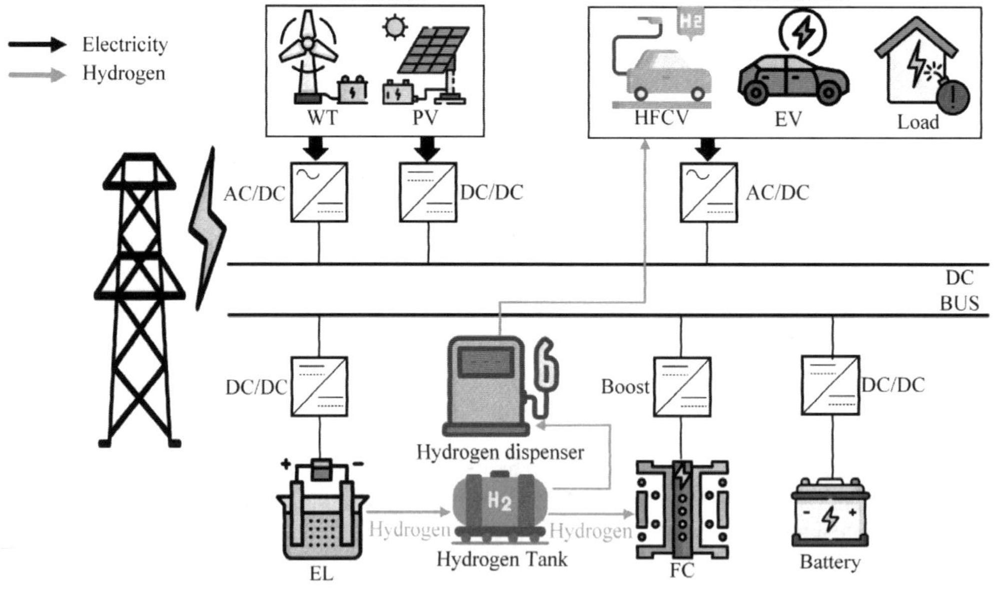

Figure 1. System configuration.

2.2. *Mathematical modeling*

The photovoltaic and wind power in this paper are obtained based on data-driven methods, so the mathematical modeling of photovoltaic and wind power will not be discussed in this section. In this section, mathematical models such as electrolytic cells and fuel cells will be introduced.

2.2.1 PEMEL modeling. In this system, a proton membrane electrolysis cell is used, and its mathematical model is as follows [6]:

$$P_t^{EL} = \eta_{EL} P_t^{Ref,EL} \tag{1}$$

$$\eta_{EL} = \eta_{EL,V} \times \eta_{EL,I} \tag{2}$$

where $\eta_{EL,V}$, $\eta_{EL,I}$, and η_{EL} respectively represent the voltage efficiency, current efficiency, and electrolytic cell efficiency of the electrolytic cell. It is the reference voltage for hydrogen fuel cells.

2.2.2 PEMFC modeling. Similar to proton membrane electrolyzers, the mathematical modeling of fuel cells is as follows [7]:

$$P_t^{FC} = \eta_{FC} P_t^{Ref,FC} \tag{3}$$

$$\eta_{FC} = \eta_{FC,V} \times \eta_{FC,I} \tag{4}$$

2.2.3 Energy storage system modeling. This paper considers using lithium battery modules and hydrogen storage tanks as energy storage systems. The hydrogen generated by the electrolytic cell is charged into the hydrogen storage tank, which can not only refuel hydrogen fuel cell vehicles but also add fuel cells for additional charging. The mathematical modeling of the energy storage system is as follows:

$$\mathrm{SOC}_t^{ES} = \mathrm{SOC}_{t-\Delta t}^{ES} + \frac{\eta^{ES.ch} P_t^{ES,ch} \Delta t}{\mathrm{Cap}^{ES}} (1 - u_t^{ES}) - \frac{P_t^{ES,dis} \Delta t}{\eta^{ES.dis} \mathrm{Cap}^{ES}} u_t^{ES} \tag{5}$$

$$\mathrm{SOH}_t^{HS} = \mathrm{SOH}_{t-\Delta t}^{HS} + \frac{P_t^{EL} \Delta t}{\mathrm{Cap}^{HS}} (1 - u_t^{HS}) - \frac{P_t^{FC} \Delta t}{\mathrm{Cap}^{HS}} u_t^{HS} - \sum_{i=0}^{u} \mathrm{H}_{t,i}^{FCEV} \sigma_{FCEV}(h_i) \tag{6}$$

where (5) represents the state of charge of the lithium battery module and (6) represents the hydrogen storage state of the hydrogen storage tank.

2.2.4 External power grid modeling. The external grid provides complementary power compensation for photovoltaic-wind systems while enabling bidirectional energy exchange for revenue generation through surplus electricity feedback, thereby achieving economic optimization. Grid-microgrid power interactions are constrained by transmission capacity limits, with the mathematical formulation expressed as:

$$P_t^{Grid} = u_t^{buy} P_{gird,t}^{buy} + u_t^{sell} P_{gird,t}^{sell} \tag{7}$$

where $P_{gird,t}^{buy}$ and $P_{gird,t}^{sell}$ respectively represent the power of electricity purchased and sold by the microgrid from the external power grid, u_t^{buy} and u_t^{sell} are logical variables to prevent the external power grid from purchasing and selling electricity simultaneously.

2.2.5 EVs and HVs modeling. In this paper, probability distributions are used to model the starting charging time and daily driving distance of electric vehicles and hydrogen fuel cell vehicles, as shown in the following equations:

$$\mathrm{PDF(SC)} = \begin{cases} \frac{1}{\sqrt{2\pi}\delta_{SC}} \exp\left(-\frac{(SC+24-\mu_{SC})^2}{2\delta_{SC}^2}\right), 0 \le SC \le \mu_{SC} - 12 \\ \frac{1}{\sqrt{2\pi}\delta_{SC}} \exp\left(-\frac{(SC-\mu_{SC})^2}{2\delta_{SC}^2}\right), \quad \mu_{SC} - 12 \le SC \le 24 \end{cases} \tag{8}$$

$$\mathrm{PDF(DD)} = \frac{1}{\sqrt{2\pi}\delta_{DD} DD} \exp\left(-\frac{(\ln DD - \mu_{DD})^2}{2\delta_{DD}^2}\right) \tag{9}$$

Considering the daily distance traveled by electric vehicles, the SOC or SOH at the charging/hydrogen refueling station can be expressed as follows:

$$\mathrm{SOC/SOH} = \frac{C_{EV/HV} - dP_{KM}}{C_{EV/HV}} \tag{10}$$

This study employs a hybrid approach combining Monte Carlo simulation and cluster analysis to construct and optimize the daily load profiles of electric vehicles and fuel cell vehicles.

3. Methodology

3.1. TCN-GRU-Attention prediction module

The TCN-GRU-Attention model integrates the multi-scale temporal modeling capability of TCN (capturing long-term periodic fluctuations), the short-term dynamic response characteristics of GRU (adapting to instantaneous mutations), and the precise focus of the attention mechanism on key features. Through parallel computing and lightweight architecture design, it ensures prediction accuracy while significantly improving computational efficiency, balancing data noise robustness and engineering deployment feasibility, and significantly optimizing the real-time stability of wind and solar load forecasting. The system architecture diagram of the TCN-GRU-Attention model is shown in Figure 2. Firstly, the convolutional causal layer and residual connection of TCN are used to output the feature matrix to the GRU layer for temporal refinement and feature extraction. Finally, the attention mechanism is used for feature fusion and prediction output.

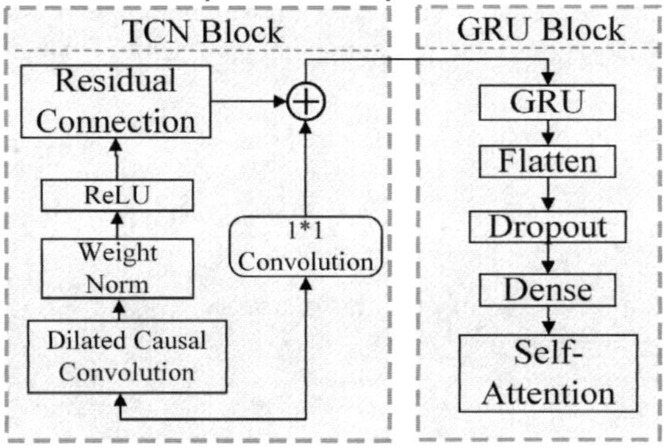

Figure 2. TCN-GRU-Attention system architecture.

3.2. Day-ahead scheduling

$$\begin{cases} \min \sum_{t=1}^{T}(C_t^{ES} + C_t^{Grid} + C_t^{Cur,WT} + C_t^{Cur,PV} + C_t^{HT} + C_t^{FC} + C_t^{EL}) \\ C_t^{ES} = \sum_{t=1}^{T} c_{ES}\,(P_t^{ES,ch} + P_t^{ES,dis})\Delta t \\ C_t^{Grid} = c_{Price} P_t^{Grid} \Delta t \\ C_t^{WT} = \sum_{t=1}^{T} c_{WT} P_t^{WT} \Delta t \\ C_t^{PV} = \sum_{t=1}^{T} c_{PV}\, P_t^{PV} \Delta t \\ C_t^{HT} = \sum_{t=1}^{T} c_{HT}\,(P_t^{HT,ch} + P_t^{HT,dis})\Delta t \\ C_t^{FC} = \sum_{t=1}^{T} c_{o\&m,FC}\, \Delta t \\ C_t^{EL} = \sum_{t=1}^{T} c_{o\&m,EL}\, \Delta t \end{cases} \tag{11}$$

The mathematical formulation of the day-ahead scheduling objective function in Phase I is presented in Equation (11), which comprises four principal cost components: the operational expenditure of the ESS C_t^{ESS} and C_t^{HT}, grid electricity procurement costs C_t^{Grid}, renewable energy operation and maintenance (O&M) costs C_t^{PV} and C_t^{WT}, and combined operational costs for electrolyzer and fuel cell systems C_t^{EL} and C_t^{FC}.

Journal of Physics: Conference Series **3031** (2025) 012001 doi:10.1088/1742-6596/3031/1/012001

3.3. Constraints

The power balance of the system is shown in Equation (12). The hydrogen power balance is shown in Equation (13), which considers the hydrogen refueling amount of the hydrogen fuel cell vehicle at time *t*.

$$P_t^{Grid} + P_t^{WT} + P_t^{PV} + (1 - u_t^{HS})P_t^{FC} + P_t^{ES,dis} = P_t^{Load} + u_t^{HS}P_t^{EL} + P_t^{ES,ch} + P_t^{EV} \tag{12}$$

$$(1 - u_t^{HS})N_t^{EL} + N_t^{HT} = u_t^{HS}N_t^{FC} + \sum_i^N H_{t,i}^{FCEV}\sigma_{FCEV}(h_i) \tag{13}$$

The constraints of lithium battery modules are shown in Equations (14)-(19), which consider the maximum and minimum SOC limits of lithium batteries, as well as charge and discharge limits.

$$0 \le P_t^{ES,ch} \le (1 - u_t^{ES})P_{t,max}^{ES,ch} \tag{14}$$

$$0 \le P_t^{ES,dis} \le u_t^{ES}P_{t,max}^{ES,dis} \tag{15}$$

$$SOC^{ES,min} \le SOC_t^{ES} \le SOC^{ES,max} \tag{16}$$

The restrictions on hydrogen storage tanks are similar to those on lithium battery modules, which also limit the maximum and minimum SOH as well as the input and output power of hydrogen storage tanks

$$0 \le P_t^{EL} \le (1 - u_t^{HS})P_{t,min}^{EL} \tag{17}$$

$$0 \le P_t^{FL} \le u_t^{HS}P_{t,max}^{FC} \tag{18}$$

$$SOH^{HS,min} \le SOC_t^{HS} \le SOC^{HS,max} \tag{19}$$

4. Result and discussion

In this section, we conducted a case study on the virtual microgrid described in Figure 1 to validate the feasibility of the proposed method, which was simulated and verified by using Gurobi solver on Matlab on a Windows PC desktop computer, equipped with an Intel Core i7-9750H CPU and 16 GB RAM.

To comprehensively assess the operational characteristics of the power system under varying supply-demand conditions, this study constructs four representative scenarios based on the two-dimensional combination of power generation and load demand. The scenarios are defined as follows:

Case 1: High wind and solar generation coupled with high electricity load.
Case 2: High wind and solar generation coupled with low electricity load.
Case 3: Low wind and solar generation coupled with high electricity load.
Case 4: Low wind and solar generation coupled with low electricity load.

The specific parameter settings simulated in this paper are as follows. PV: efficiency 0.2, area 500 m^2; Lithium battery: charge-discharge efficiency 0.95, max capacity 100 kW, max charge-discharge power 20 kW; Hydrogen storage tank: charge-discharge efficiency 0.9, max energy storage 1 m^3, max charge-discharge power 0.25 m^3; PEMEL: max electrolysis power 30 kW, conversion efficiency 0.95; PEMFC: conversion efficiency 0.95, max power 35 kW.

4.1. Comparison of prediction results

To validate the superiority of the proposed model, the MAE was employed as a quantitative metric for assessing prediction accuracy. Furthermore, benchmark analyses were performed against conventional time-series forecasting algorithms, including LSTM and GRU networks, to demonstrate the comparative advantages of the proposed methodology. Figures 3-5 present a multi-temporal resolution comparative analysis between predicted and empirically measured generation profiles.

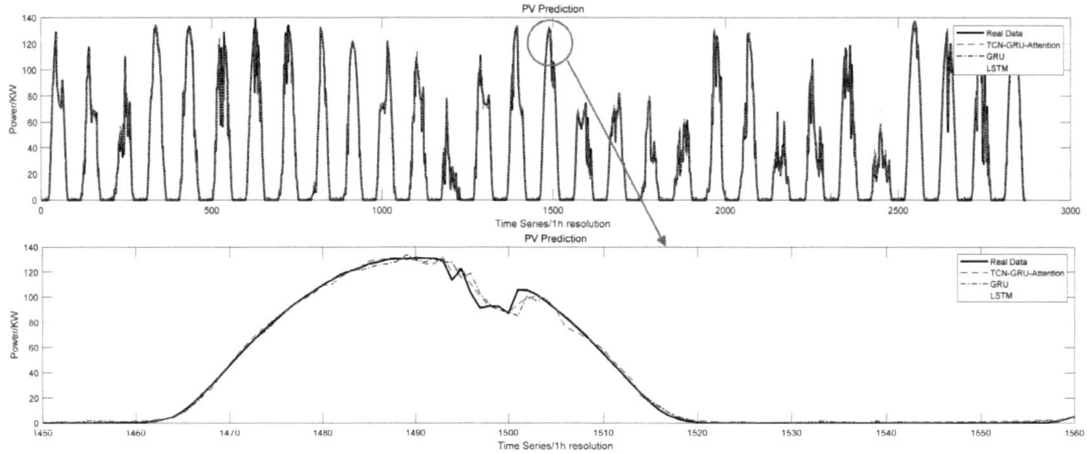

Figure 3. PV prediction of different prediction methods.

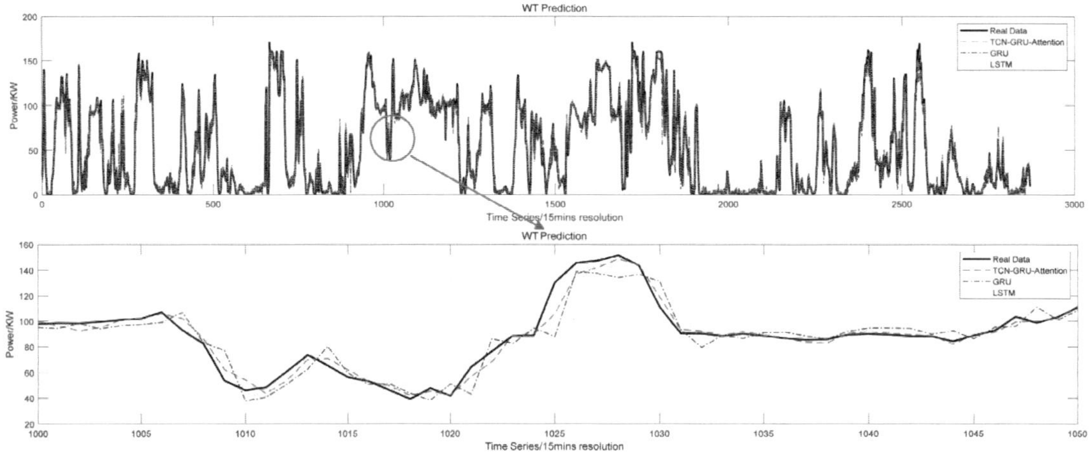

Figure 4. WT prediction of different prediction methods.

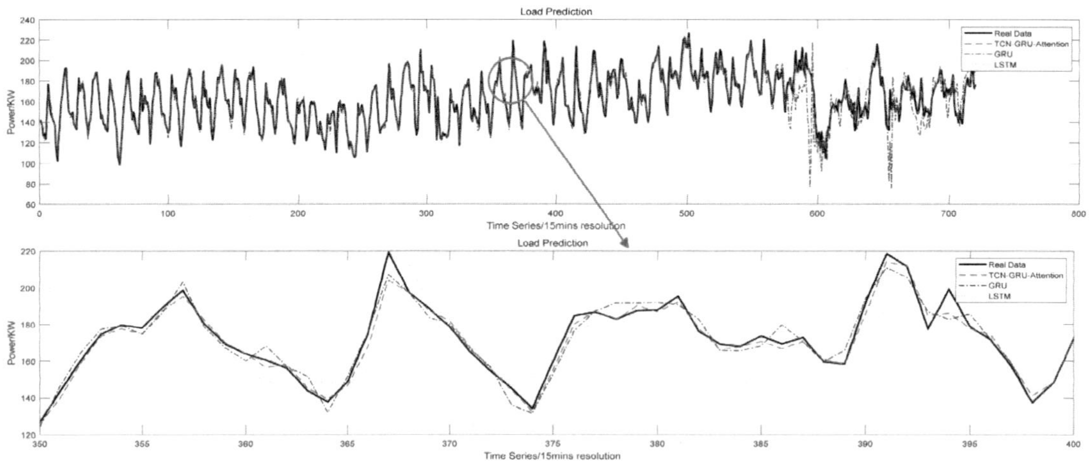

Figure 5. Load prediction of different prediction methods.

As delineated in Table 1, the PV generation predictions demonstrated superior predictive accuracy, followed by load demand forecasts, while wind power predictions exhibited the highest MAE due to the inherent stochasticity of wind speed dynamics. The proposed forecasting methodology achieved significantly lower MAE values across all prediction categories compared to benchmark models, evidencing its enhanced capability in renewable generation and load demand forecasting precision.

Table 1. Comparative MAE Performance of Forecasting Algorithms for Renewable Generation and Load Demand.

	Photovoltaic power generation	Wind power generation	Load power generation
TCN-GRU-Attention	1.771	2.992	2.836
LSTM	4.382	6.06	9.474
GRU	3.427	5.621	6.273

4.2. New energy vehicle load generation results

As illustrated in Figure 6, the load demand curves of new energy vehicles under distinct scenarios are depicted, with the blue line representing the low-load scenario and the red line indicating the high-load scenario. The horizontal axis denotes the temporal resolution, while the vertical axis quantifies normalized load power. The figure reveals that both high-load and low-load scenarios reach peak demand intensities around 18:00, a pattern likely driven by synchronized charging activities of vehicle fleets returning from daily commutes. During this period, the high-load scenario exhibits a 1.1–1.4-fold surge in peak magnitude compared to low-load conditions, reflecting the compounding effects of aggregated large-scale charging events that amplify grid stress through temporal concentration of power demand.

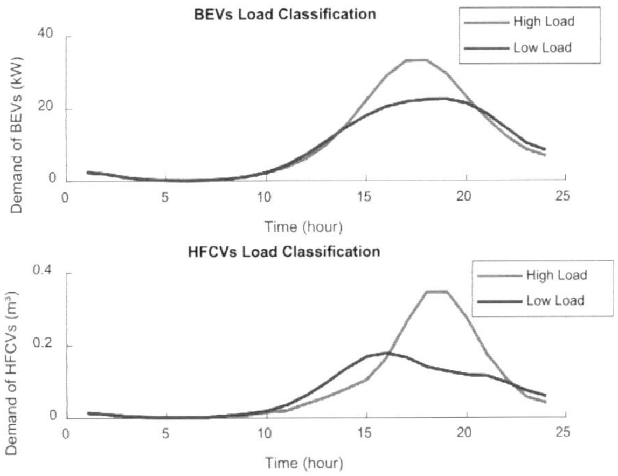

Figure 6. Comparison of load curves for electric vehicles and hydrogen fuel cell vehicles.

4.3. Day-ahead scheduling result

Figure 7 shows distinct grid dispatch patterns in four scenarios defined by distributed energy resource (DER) generation levels and load demand. In Case 1 (high DER, high load), the system is nearly self-sufficient, relying little on the external grid due to the strong match between local generation and demand. Energy storage systems balance short-term power mismatches. Case 2 (high DER, low load) has net reverse power flow to the grid. As the local load cannot absorb the surplus DER power, overvoltage risks arise. Curtailing excess generation or using energy storage is needed to mitigate this. Case 3 (low DER, high load) depends heavily on grid imports as DER cannot meet demand peaks, highlighting the need for capacity reserves and demand-side flexibility mechanisms. Case 4 (low DER, low load) has a stable yet suboptimal operation, with minimal grid interaction and DER underutilization because of reduced supply and demand.

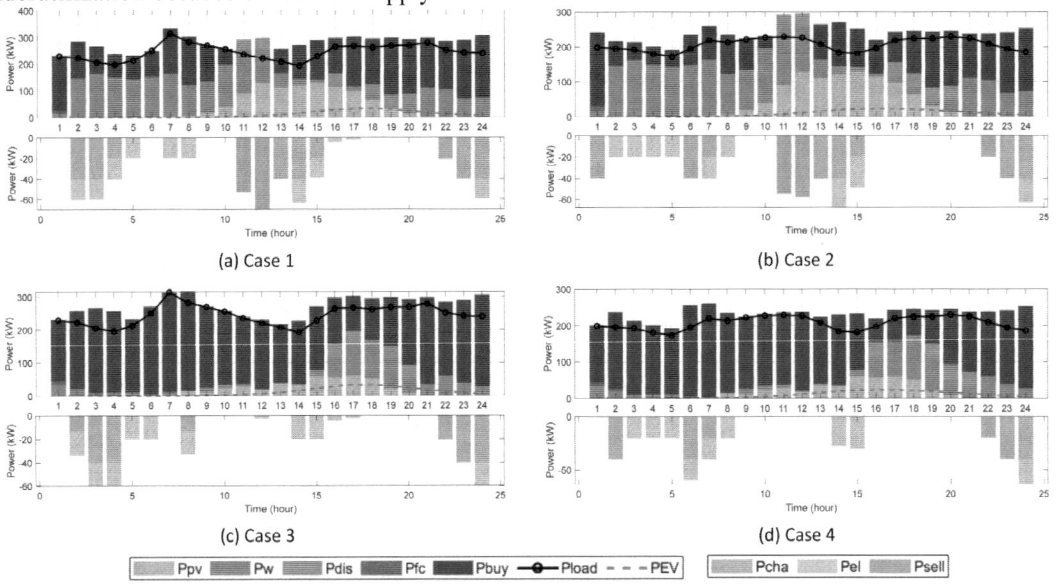

Figure 7. Output of each piece of equipment in the day-ahead scheduling phase.

4.4. Day-ahead scheduling result

Table 2 shows the scheduling economic costs of different algorithms in four scenarios. The algorithm proposed in this paper, with its high prediction accuracy, achieves a more economical scheduling scheme. For extreme cases like Case 3 (low generation/high load), accurate source-load prediction significantly impacts microgrid economic dispatch. When renewable energy generation is low, the external grid provides more power, resulting in a relatively high cost in Case 4. In summary, the proposed algorithm exhibits economic benefits in normal operation and robustness in extreme situations. By accurately predicting generation and demand, it optimizes micro-grid energy allocation, reducing external grid dependence and overall costs.

Table 2. Operating Costs under Different Scenarios and Algorithms.

	Proposed method	GRU	LSTM	Without forecast
Case 1	2650.74	2693.83	2714.62	2735.31
Case 2	1890.02	1934.14	1946.57	1893.17
Case 3	3917.82	3932.96	3979.42	4130.87
Case 4	3156.2	3169.25	3195.93	3205.71

5. Conclusion

This paper proposes a day-ahead scheduling optimization framework for hydrogen-electric hybrid microgrids to address operational challenges caused by multi-domain uncertainties from supply-load interactions. Firstly, multi-physics coupled models of electrolyzers and fuel cells were developed. Subsequently, we combine TCN employing dilated causal convolutions for meteorological long-term pattern mining with GRU featuring adaptive gating mechanisms to mitigate renewable transient fluctuations, enhanced by self-attention-based feature fusion layers for coordinated generation-load prediction. To address the dual uncertainties arising from significant stochasticity in electric vehicles and subjective behavioral patterns of vehicle owners, this study establishes a probabilistic density behavioral function for users and generates charging characteristic curves with load features through Monte Carlo-based scenario simulations. Simulation results and irradiation/wind speed datasets demonstrate that the proposed TCN-GRU hybrid model reduces MAE by 34.8% compared to LSTM and Transformer baselines while scheduling achieves 5.1-5.4% operational cost savings over conventional deterministic strategies substantiating long-term economic and stability enhancements.

Acknowledgments

This work was financially supported by the National Key Research and Development Program of China (2020YFB1506802).

References

[1] Lasseter, R. H. (2002) 'MicroGrids', in 2002 IEEE Power Engineering Society Winter Meeting. Conference Proceedings (Cat. No. 02CH37309). Winter Meeting of the Power Engineering Society, New York, NY, USA: IEEE, pp. 305–308. Available at: https://doi.org/10.1109/PESW.2002.985003.
[2] Li, J. et al. (2022) 'Towards net-zero smart system: An power synergy management approach of hydrogen and battery hybrid system with hydrogen safety consideration', Energy Conversion and Management, 263, p. 115717. Available at: https://doi.org/10.1016/j.enconman.2022.115717.
[3] Ceylan, C. and Devrim, Y. (2023) 'Green hydrogen-based off-grid and on-grid hybrid energy systems', International Journal of Hydrogen Energy, 48(99), pp. 39084–39096. Available at: https://doi.org/10.1016/j.ijhydene.2023.02.031.
[4] Marocco, P. et al. (2021) 'Optimal design of stand-alone solutions based on RES + hydrogen storage feeding off-grid communities', Energy Conversion and Management, 238, p. 114147. Available at: https://doi.org/10.1016/j.enconman.2021.114147.
[5] Zhang, Y., Meng, F., Wang, R., Zhu, W., & Zeng, X. J. (2018). 'A stochastic MPC based approach to integrated energy management in microgrids', Sustainable Cities and Society, 41, 349–362. https://doi.org/10.1016/j.scs.2018.05.044.
[6] Görgün, H. 'Dynamic modeling of a proton exchange membrane (PEM) electrolyzer', International Journal of Hydrogen Energy 31.1 (2006): 29–38.
[7] Zakeri, B., and Sanna, S. 'Electrical energy storage systems: A comparative life cycle cost analysis', Renewable and Sustainable Energy Reviews 42 (2015): 569–596.

Journal of Physics: Conference Series **3031** (2025) 012002 doi:10.1088/1742-6596/3031/1/012002

Design and research on the hydrogen liquefaction process coupled with LNG precooling and multistage deep cooling

Xinglin Yang*, Min Zhou

School of Energy and Power, Jiangsu University of Science and Technology, Zhenjiang 212003, China

*Corresponding author's e-mail: hcyangxl2010@163.com

Abstract. A novel hydrogen liquefaction process is proposed to address the high energy consumption issue in hydrogen liquefaction processes. Based on the dual mixed-refrigerant hydrogen liquefaction process, a new LNG-precooled hydrogen liquefaction process is introduced. This process integrates an LNG cold energy precooling system with a cascade refrigeration-assisted precooling system. Using Aspen HYSYS software, simulations and optimizations were conducted with the specific energy consumption (*SEC*) as the objective function. The results demonstrate that the proposed process achieves an *SEC* of 4.49 kWh/kgLH$_2$ and an exergy efficiency of 64.05%. The exergy losses in the ortho-para hydrogen converters and expanders are the most significant, accounting for 39.91% and 31.64%, respectively.

1. Introduction

Hydrogen, as the most abundant energy carrier on Earth, offers advantages such as high calorific value (approximately 35 kWh/kgLH$_2$), zero emissions, and high energy density per unit volume, making it a highly anticipated renewable and clean energy source for the 21st century [1]. As a widely available and low-carbon secondary energy source [2], hydrogen plays a critical role in the global energy transition and has broad application prospects [3]. However, liquefaction requires substantial electrical energy [4]. Recent studies report specific energy consumption (*SEC*) values of hydrogen liquefaction systems ranging from 10~12 kWh/kgLH$_2$, with efficiencies between 25% and 40%, where 36% of available energy is consumed per kilogram of liquefied hydrogen [5]. Thus, reducing energy consumption in hydrogen liquefaction is a crucial research focus.

In the field of nitrogen-precooled hydrogen liquefaction, Baker et al. [6] proposed a dual-pressure Claude hydrogen liquefaction process with an *SEC* of 10.85 kWh/kgLH$_2$ and an exergy efficiency of 36%. Zhu et al. [7] introduced a hydrogen liquefaction process combining mixed-refrigerant precooling and hydrogen expansion deep cooling. When the mixed refrigerant composition was methane (10.2%), ethane (29.3%), propane (39.3%), and nitrogen (21.2%), the precooling cycle achieved the minimum specific power consumption, while the deep cooling cycle reduced specific power consumption by 7.8% compared to the initial parameters.

To further reduce energy consumption, this paper proposes an LNG-precooled and multistage deep-cooling coupled hydrogen liquefaction process. Simulations and optimizations of the process were conducted using Aspen HYSYS software to analyze its *SEC*, efficiency, and heat exchange characteristics, thereby verifying its feasibility.

2. LNG-precooled and multistage deep-cooling coupled hydrogen liquefaction process

2.1. Process design

Building on the cascade refrigeration-assisted precooling hydrogen liquefaction process proposed by Shen et al. [8]. This study introduces a modified process. Due to the high energy efficiency of mixed refrigerant cooling[9], the new process retains the mixed-refrigerant deep cooling system where the cryogenic section remains as a mixed-refrigerant cycle, but replaces the complex precooling section with an integrated LNG cold energy precooling and cascade refrigeration-assisted precooling system, resulting in a simplified hydrogen liquefaction system (Figure 1). The system comprises a hydrogen liquefaction system, an NH_3/CO_2 cascade refrigeration system, a precooling system, and a mixed-refrigerant deep cooling system.

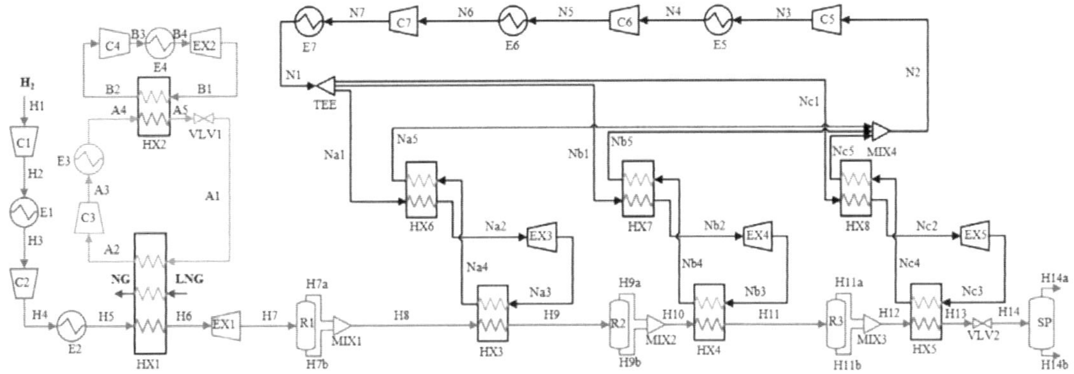

Figure 1. LNG-precooled and multistage deep-cooling coupled with a hydrogen liquefaction process flow.

The feed hydrogen sequentially passes through compressor C1, cooler E1, compressor C2, and cooler E2 for two-stage compression and cooling. It is then further cooled via heat exchanger HX1 and expander EX1 before entering the primary ortho-para hydrogen converter R1 for adiabatic conversion. After temperature elevation, the hydrogen is cooled in heat exchanger HX3 and subsequently undergoes stepwise cooling and conversion through the secondary ortho-para hydrogen converter R2, heat exchanger HX4, tertiary ortho-para hydrogen converter R3, and heat exchanger HX5. Finally, the cryogenic hydrogen is liquefied via expansion in throttle valve VLV2, and the liquefied hydrogen is stored in a tank.

NH_3/CO_2 cascade refrigeration system: The high-temperature refrigeration cycle (CO_2) is compressed by compressor C4, cooled in cooler E4, then expanded through expander EX2 and evaporated in heat exchanger HX2 to cool the NH_3 in the low-temperature cycle. In the low-temperature cycle, NH_3 is compressed by compressor C3, cooled in cooler E3, exchanges heat in HX2, depressurized via throttle valve VLV1, and finally enters HX1 to precool the feed hydrogen.

The precooling system employs a combined method integrating LNG with NH_3/CO_2 cascade refrigeration. Both the LNG and cascade systems simultaneously supply the required temperature and cooling capacity for hydrogen precooling in heat exchanger HX1, followed by further cooling via expander EX1.

The mixed-refrigerant deep cooling system employs a blend of H_2, He, and Ne as refrigerants. Building on the cryogenic system described in Reference [10], this process incorporates an additional compression-cooling stage, resulting in a total of three compression-cooling stages. Specifically, the mixed refrigerant sequentially passes through compressor C5, cooler E5, compressor C6, cooler E6, compressor C7, and cooler E7. It then undergoes temperature reduction via self-cooling heat exchangers HX6, HX7, and HX8, followed by further cooling through expanders EX3, EX4, and EX5. Finally, the refrigerant cools hydrogen to -253 °C after heat exchangers HX3, HX4, and HX5.

2.2. Process simulation

The mixed-refrigerant hydrogen liquefaction process described above was modeled and subjected to steady-state simulation using Aspen HYSYS software, with the Peng-Robinson equation applied in the simulation. The simulation conditions are as follows:

- Hydrogen liquefaction process capacity: 9.85 kg/s.
- Feed hydrogen conditions: Pure hydrogen at 25°C and 2100 kPa.

- Pressure drops in multi-stream heat exchangers and coolers: 0 kPa.
- Compressor adiabatic efficiency: 90%; expander isentropic efficiency: 85%.
- Minimum temperature difference in multi-stream heat exchangers: 1~3°C.
- Liquid hydrogen (LH_2) product specifications: Pressure of 130 kPa, para-hydrogen concentration ≥ 95%.

The three-stage ortho-para hydrogen converters in the simulation model were simplified based on the exothermic reaction characteristics of the conversion process, represented as p-Hydrogen + Heat, with conversion rates modeled as temperature-dependent functions:

$$Conversion(\%) \rightarrow C_0 + C_1 \times T + C_2 \times T^2 \tag{1}$$

C_1, C_2: conversion coefficient;
T: hydrogen temperature, K.

Coefficients for each converter (R1-R3) are listed in Table 1.

Table 1. Conversion coefficients for ortho-para hydrogen converters [11].

Ortho-para hydrogen converter	Conversion coefficient		
	C_0	C_1	C_2
R1	44.45	-0.04578	1.158e-003
R2	73.19	-0.04578	1.158e-003
R3	96.52	-0.04578	1.158e-003

2.3. *System optimization and evaluation*

The system was optimized using the Original optimizer in Aspen HYSYS software, with the *SEC* set as the objective function. Twenty parameters significantly affecting system performance - including compressor outlet pressures, expander outlet pressures, and refrigerant mass flow rates - were selected as decision variables. The *SEC* of the hydrogen liquefaction process is calculated as:

$$SEC = (W_{\mathrm{Com}} - W_{\mathrm{Exp}}) / m_{\mathrm{LH_2}} \tag{2}$$

SEC: Specific energy consumption, kWh/kg;
W_{Com}: Total energy consumed by compressors, kW;
W_{Exp}: Energy recovered by expanders, kW;
$m_{\mathrm{LH_2}}$: Mass flow rate of liquefied hydrogen, kg/h.

In addition to the *SEC*, this study employs exergy efficiency and exergy loss to evaluate the thermodynamic perfection of the proposed process. Exergy represents the maximum useful work obtainable from a system when it reaches equilibrium with the environment. The exergy efficiency (E_x) is defined as the ratio of the theoretical minimum liquefaction work to the actual work consumption. A higher E_x indicates lower energy losses in the system and improved economic viability.

3. Results and analysis

3.1. *System performance optimization results and analysis*

Optimized decision variables and post-optimization parameters are summarized in Table 2. The optimized *SEC* is 4.49 kWh/kgLH_2, with unit energy consumption of 0.763 kWh/kgLH_2 from the precooling system and 0.117 kWh/kgLH_2 from the cascade refrigeration system. The exergy efficiency reaches 64.05%. Compared to the original mixed-refrigerant precooled hydrogen liquefaction process proposed by Shen et al. [8], the *SEC* is reduced by 35.2%, and the exergy efficiency is improved by 9.81%.

Table 2. Optimization variables and results.

Optimization variables	Lower limit	Optimized value	Upper limit
Compressor C1 outlet pressure /kPa	4000.00	5423.17	7000.00
Compressor C2 outlet pressure /kPa	10000.00	13354.28	16000.00
Compressor C3 outlet pressure /kPa	60.00	62.01	180.00
Compressor C4 outlet pressure /kPa	45.00	86.83	170.00
Compressor C6 outlet pressure /kPa	200.00	261.01	350.00
Compressor C7 outlet pressure /kPa	360.00	483.41	580.00
Compressor C8 outlet pressure /kPa	590.00	911.23	1400.00
Expander EX1 outlet pressure /kPa	1000.00	1900.03	3000.00
Expander EX2 outlet pressure /kPa	20.00	28.23	40.00
Expander EX3 outlet pressure /kPa	100.00	130.30	240.00
Expander EX4 outlet pressure /kPa	100.00	166.67	240.00
Expander EX5 outlet pressure /kPa	100.00	163.05	240.00
NH_3 mass flow rate /(kg/h)	40000.00	57208.65	60000.00
CO_2 mass flow rate /(kg/h)	100000.00	115447.48	150000.00
N1 mass flow rate /(kg/h)	120000.00	155524.20	180000.00
Na1 mass flow rate /(kg/h)	16200.00	34674.20	60000.00
Nc1 mass flow rate /(kg/h)	34500.00	58609.97	119000.00
Outlet temperature of Na1 after passing through the self-cooling heat exchanger /°C	-186.00	-184.02	-182.00
Outlet temperature of Nb1 after passing through the self-cooling heat exchanger /°C	-235.00	-224.56	-220.00
Outlet temperature of Nc1 after passing through the self-cooling heat exchanger /°C	-253.00	-247.53	-247.00

Figure 2 illustrates the distribution of exergy losses across major system components. The exergy losses of compression equipment, heat exchange equipment, expansion equipment, cooling equipment, ortho-para hydrogen converter, and throttling equipment are 4518 kW, 8787 kW, 24900 kW, 7603 kW, 31420 kW, and 1494 kW. The total exergy loss amounts to 78722 kW. As shown in the figure, the ortho-para hydrogen converters and expanders account for the largest shares of exergy loss at 39.91% and 31.64%, respectively. Additionally, the exergy destruction in heat exchangers represents 11.16% of the total exergy loss.

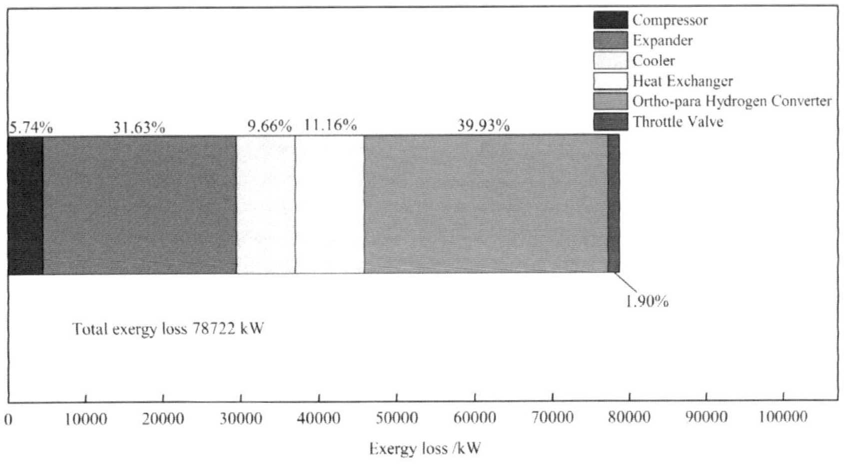

Figure 2. Exergy loss accumulation diagram of the main equipment during hydrogen liquefaction.

Journal of Physics: Conference Series **3031** (2025) 012002 doi:10.1088/1742-6596/3031/1/012002

3.2. Heat exchange composite curve analysis

The composite curve is a thermodynamic graphical tool widely used to evaluate process efficiency. In heat exchangers, the composite curves of hot and cold streams should be positioned as close as possible to achieve low energy consumption and high efficiency. Figure 3 illustrates the optimized composite curves of hot and cold streams in the pre-cooling and cryogenic heat exchangers of the system. To ensure the simulation aligns with real-world conditions, a minimum temperature difference of ≥ 1°C must be maintained during the heat exchange process.

After optimization by the optimizer, the composite curves of hot and cold streams within the heat exchangers demonstrate good overall alignment, indicating efficient heat exchange performance in the process system. However, significant spacing remains in HX1, particularly in the -105°C to 25°C range, suggesting severe heat loss in this exchanger. This is attributed to the large temperature span of the cold energy exchange in HX1, highlighting its potential for further optimization.

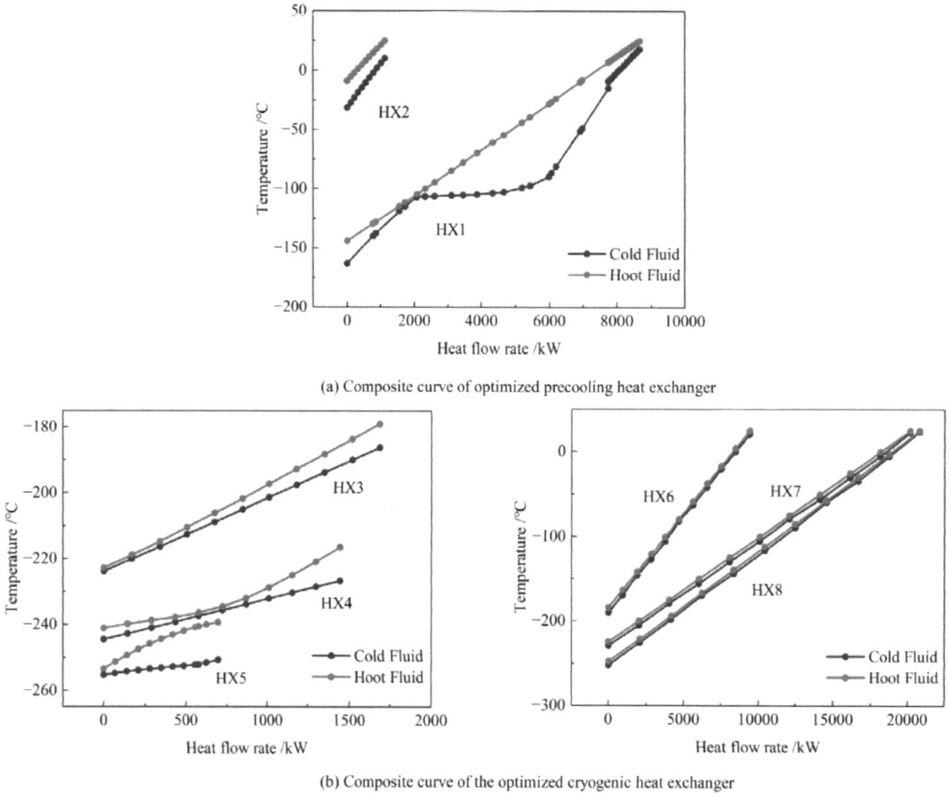

Figure 3. Composite curves of precooling and cryogenic heat exchangers after optimization.

4. Conclusion

Based on the results and analysis presented above, the following conclusions are obtained as below:

(1) Based on existing dual mixed refrigerant hydrogen liquefaction processes, this study proposes a novel hydrogen liquefaction process featuring simplified configuration, low energy consumption, high efficiency, and LNG pre-cooling. The integration of LNG cold energy pre-cooling with cascade refrigeration-assisted pre-cooling in the pre-cooling system significantly enhances process performance. The proposed process achieves a specific energy consumption (*SEC*) of 4.49 kWh/kgLH$_2$ and an exergy efficiency of 64.05%, outperforming hydrogen liquefaction processes reported in relevant literature.

(2) The ortho-para hydrogen converter and expanders contribute dominantly to the total exergy losses in the system.

(3) In this process, the composite curves of hot and cold streams exhibit good alignment in most heat exchangers, indicating efficient heat transfer and minimal energy losses. However, HX1 shows suboptimal composite curve matching due to its wide temperature span in cold energy exchange, suggesting potential for further optimization.

References

[1] Hammad A, Dincer I (2018) Analysis and assessment of an advanced hydrogen liquefaction system. International Journal of Hydrogen Energy, 43 (2): 1139-1151.

[2] Zhou Y X, Zhang Z Y, Liu H T, Feng C (2025) Current situation and development trend of large-scale hydrogen production technology from renewable energy. Low Temperature and Specialty Gases, 43 (01): 15-18+33.

[3] Fu G Y (2019) Hydrogen energy in China's energy transition: Role and strategic significance. China Coal, 45 (10): 15-21.

[4] Park S, Hashim B, Zahid U, Kim J (2025) Global risk assessment of hydrogen refueling stations: Trends, challenges, and future directions. International Journal of Hydrogen Energy, 106: 1462-1479.

[5] Qyyum M A, Riaz A, Naquash A, et al. (2021) 100% saturated liquid hydrogen production: Mixed - refrigerant cascaded process with two - stage ortho - to - para hydrogen conversion. Energy Conversion and Management, 246: 114659-.

[6] Baker C R, Shaner R L (1978) A study of the efficiency of hydrogen liquefaction. International Journal of Hydrogen Energy, 3 (3): 321-334.

[7] Zhu J Q, Yang K, Xiong L Y (2024) Optimization and process analysis of hydrogen liquefaction process for precooling of mixed working medium. Cryogenics and Superconductivity, 52 (07): 75-82+88.

[8] Shen J B, Luo L M, Jiang L L, Jiang Q F, Guo T (2022) Design and study of hydrogen liquefaction process auxiliary precooling by cascade refrigeration. Cryogenics and Superconductivity, 50 (11): 99-104.

[9] Liu L, Hu Y R, Huang J L, et al. (2025) Comparison and optimization of mixed refrigerants hydrogen liquefaction process. Modern Chemical Industry, 45 (03): 227-232.

[10] Sadaghiani M S, Mehrpooya M (2017). Introducing an energy analysis of a novel cryogenic hydrogen liquefaction process configuration. International Journal of Hydrogen Energy, 42 (9): 6033-6050.

[11] Asadnia M, Mehrpooya M (2017) A novel hydrogen liquefaction process configuration with combined mixed refrigerant systems. International Journal of Hydrogen Energy, 42 (23): 15564-15585.

Journal of Physics: Conference Series **3031** (2025) 012003 doi:10.1088/1742-6596/3031/1/012003

Sensitivity analysis of critical design parameters for insulating compressed air energy storage systems

Zhirong Jiang[1*], Zhongbo Hu[1], Jianchuan Wu[1], Huan Zhou[1], Qing He[2], Xu Liu[2]

[1]POWERCHINA Chengdu Engineering Corporation Limited, Chengdu 611100, China
[2] School of Energy Power and Mechanical Engineering, North China Electric Power University, Beijing, 102206, China

[*]Corresponding author's e-mail: 1475018970@qq.com

Abstract: The study establishes a 100 MW insulating air pressure energy storage system by Ebsilon software. An in-depth analysis has been conducted on how variations in design parameters affect the thermal performance of the compressed air energy storage (CAES) system. The results show the following: Increasing the compression ratio has two effects on the system. There is an obvious decline of mass flow when the compression ratio rises, while system efficiency increases with an increase in the compression ratio; When the expansion ratio increases, the air mass within the system decreases from 251.13 kg/s to 234.24 kg/s upon a consistent power output, and the circulated efficiency shows an increasing trend; With the boost of storage temperature, system would operate more efficiently with certain ranges before decreasing, with a critical point at a storage temperature of approximately 222.32 °C; Rising in heat exchange temperature difference bring a decrease in system storage temperature, while the air mass flow rate increases. Therefore, under the significant influence of the climbing air mass flow rate, the system cycle efficiency illustrates a decreasing trend.

1. Introduction

As the energy structure continues to be optimized, the proportion of non-fossil fuel power installed capacity increases year by year. With the large-scale development of green energy sources such as wind and solar power, issues such as power fluctuations and large peak-valley differences become increasingly prominent, making the demand for extensive, durable energy storage urgent [1]. Adiabatic CAES is a kind of advanced technology for energy storage with huge capacity, which is also clean and durable. Therefore, in the context of the energy and electricity industry's strong development efforts, CAES technology has attracted numerous concentrations. Meanwhile, higher requirements have been placed on its further optimization and market application [2]. To analyze the energy efficiency of adiabatic compressed air energy storage (AA-CAES) systems, an efficiency analytical model for AA-CAES was established, indicating that when thermal storage performance is lower than that of compression, system efficiency rises with the additional stages; conversely, when thermal storage efficiency exceeds compression efficiency, system efficiency decreases with an increasing number of phases [3]. Through thermodynamic analysis of multi-stage compressed air energy storage systems [4], a thermodynamic model for compressed air energy storage systems was developed, and the effects of various configurations on volume flow variations in AA-CAES systems were analyzed [5-6]. An optimization algorithm based on system efficiency was proposed, and the wide operating range of compressed air energy storage systems was defined [7]. Previous studies have conducted in-depth analyses of system mechanisms or specific key parameters; however, the analysis of key parameters in engineering applications remains incomplete.

This study combines key design parameters from engineering applications, utilizing Ebsilon software to establish a 100 MW adiabatic CAES system. It deepens the exploration of how this system is affected by parameters including pressure ratio, expansion proportion, number of compression stages, final stage outlet pressure, storage temperature, and heat conveying temperature difference. The study calculates how changes in parameter values influence system efficiency and determines the underlying reasons for these effects. Furthermore, it proposes parameter selection recommendations to better inform project design planning. The technical approach is illustrated in Figure 1.

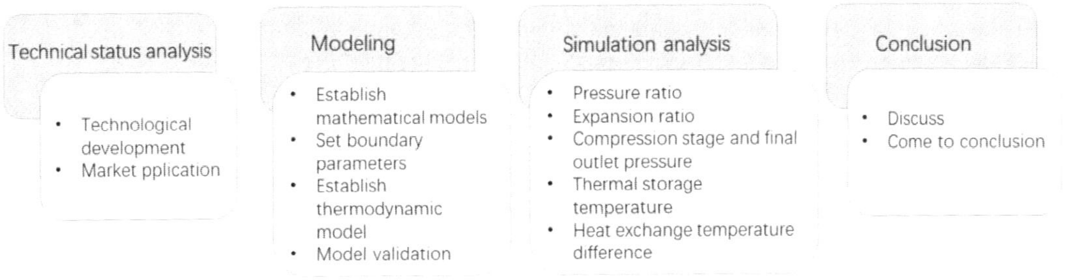

Figure 1. The technical approach of this paper.

2. System Principles

The working principle of the insulating CAES system is shown in Figure 2. Air from the environment is transferred to a high-pressure state after flowing into the compressor. It then passes through the after-compressor heat exchanger, where a thermal storage medium absorbs the heat from the high-pressure air and stores it. The cooled high-pressure air is stored in storage tanks or underground caverns. During energy release, the stored high-pressure air passes through the heat exchanger to absorb the stored heat, heating it to a high-temperature state before entering the turbine. There, the air undergoes expansion and does work, driving the generator to produce electricity [8].

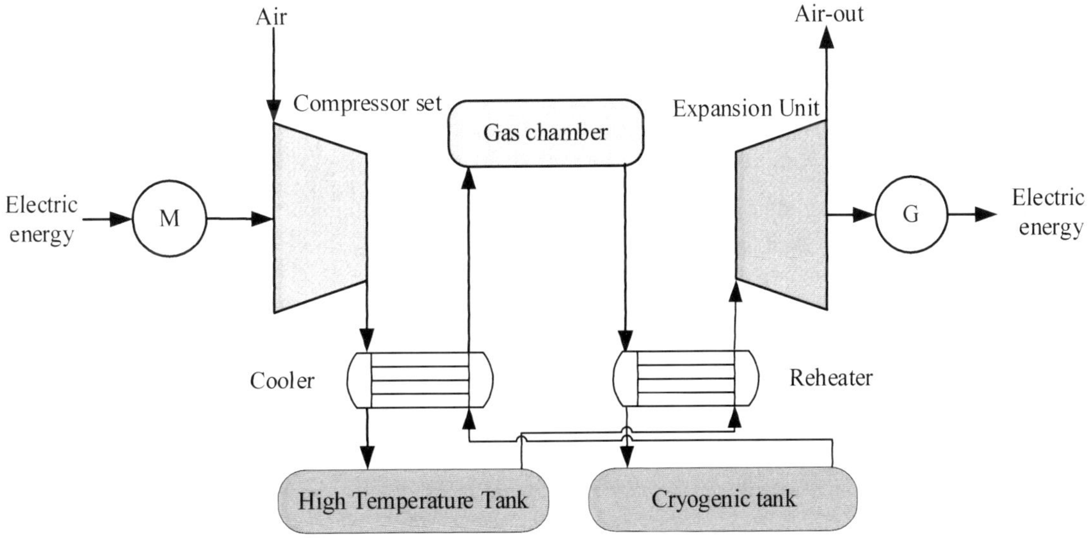

Figure 2. Adiabatic compressed air energy storage (A-CAES) system schematic diagram.

Advanced adiabatic compressed air energy storage technology typically uses water, thermal oil, or molten salt as thermal storage media. From an economic perspective, thermal oil has the highest cost, followed by molten salt storage, while pressurized water is the most economical. Therefore, this system employs pressurized water as the thermal storage medium.

Journal of Physics: Conference Series **3031** (2025) 012003 doi:10.1088/1742-6596/3031/1/012003

3. System modeling

This study conducts a survey analysis of the primary design parameter operating data from an already operational CAES power plant during its operation. It reveals that CAES systems are emerging technologies, lacking an in-depth understanding of how key parameters affect the system during the design process. Based on the research objective of this paper, we model the entire process of a 100 MW insulating CAES system. To simplify this system, we make the following hypotheses [9-10]:

1) The operating conditions of the energy storage system are steady-state.
2) Heat loss and frictional losses in various components are negligible.
3) Pressure drop and energy consumption in the pipelines are not considered.
4) Compression and expansion equipment operates under stable conditions.
5) Heat exchangers employ counterflow for complete heat exchange.
6) The sealing and thermal insulation properties of the storage tank and gas chamber are excellent.
7) We set the initial ambient temperature to 25 °C.

3.1. *Specimens results*

3.1.1. Compression system. During the simulation process, we consider the air as an ideal gas. The internal working process of the compressor is polytropic; however, due to the high flow rate and rapid flow velocity of the working fluid, the heat exchange process cannot be completed in a short time. Therefore, under normal circumstances, it can be considered an adiabatic process. Given the inlet air temperature of the compressor, the outlet air temperature is:

$$T_{co}(i)=\left[1+\frac{\beta_c(i)^{\frac{\kappa-1}{\kappa}}-1}{\eta_c(i)}\right]T_{ci}(i), i=1,2,\cdots,N \tag{1}$$

where $T_{co}(i)$ is the outlet air temperature of the compressor in Kelvin (K); $\beta_c(i)$ is the compression ratio of the compressor; $\eta_c(i)$ is the adiabatic efficiency of the compressor (%); κ is the adiabatic index of air; $T_{ci}(i)$ is the inlet air temperature of the compressor; N is the number of stages in the compressor.
The compression ratio is defined as:

$$\beta_c(i)=p_{co}(i)/p_{ci}(i) \tag{2}$$

where $p_{co}(i)$ indicates the compressor outlet air pressure, in (Pa) and $p_{ci}(i)$ is the compressor inlet air pressure, in (Pa).

The air compressor operates under stable conditions, and during the energy storage process, both the outlet air pressure and temperature of the compressor remain constant. The air compression process in the system is accomplished by a combination of multi-stage compressors and interstage coolers alternately connected in series. The total power consumption of the compressor unit P_c is:

$$P_c=m_c\sum_{i=1}^{N}c_a T_{ci}(i)\frac{\beta_c(i)^{\frac{\kappa-1}{\kappa}}-1}{\eta_c(i)} \tag{3}$$

where c_a indicates the specific air heat, in J/(kg·K); m_c is the mass flow rate of compressed air, in (kg/s).

3.1.2. Expansion power generation system. The correlation between the expander's inlet and outlet unit is:

$$T_{eo}(i)=\beta_e(i)^{\frac{\kappa-1}{\kappa}}T_{ei}(i), i=1,2,3,\cdots,M \tag{4}$$

where $T_{eo}(i)$ is the export air temperature of the expander in Kelvin (K); $\beta_e(i)$ denotes the inlet air temperature of the expander in Kelvin (K); and M represents the number of stages in the expander unit.

The expansion ratio $\beta_e(i)$ is:

$$\beta_e(i) = p_{eo}(i) / p_{ei}(i) \tag{5}$$

In Equation (5), $p_{eo}(i)$ denotes the expansion machine outlet air pressure in(Pa) and $p_{ei}(i)$ is the expansion machine inlet air pressure in (Pa).

The expander unit in the system consists of multiple stages of expanders and interstage reheaters connected alternately in series. The high-pressure air expands to do work, which is also a steady process, with the outlet pressure and temperature of each stage being constant. The high-pressure air sequentially passes through every stage of the expander to perform expansion work, and the total export energy of the expander unit P_e is:

$$P_e = m_e \sum_{i=1}^{M} \eta_e(i) c_a T_{ei}(i) \left[1 - \beta_e(i)^{\frac{\kappa-1}{\kappa}}\right] \tag{6}$$

where $\eta_e(i)$ denotes the adiabatic efficiency of the expansion machine (%) and m_e is the mass flux of air discharged from the expansion machine in (kg/s).

3.1.3. Heat exchanger. In the process of discharging energy, gas with high pressure released from the storage device has a relatively low temperature. Before driving the expander to perform work, it first undergoes heat exchange and heating through a heater to raise its temperature. Therefore, the import of the first-stage heater is equal to the export gas temperature of the valve. After flowing through the i-th stage heat exchanger, the gas temperature of the entrance becomes the inlet temperature of the i-th stage expander.

The outlet air temperature of the reheater is:

$$T_{eao}(i) = \eta_{he} T_{eoi}(i) + (1 - \eta_{he}) T_{eai}(i), i = 1, 2, \cdots, M \tag{7}$$

where $T_{eao}(i)$ is the air temperature at the outlet of the reheater in Kelvin (K); η_{he} is the heat exchange efficiency of reheater (%); $T_{eoi}(i)$ is the temperature of the heat exchange medium at the inlet of the reheater in Kelvin (K) and $T_{eai}(i)$ is the air temperature at the inlet of the reheater in Kelvin (K).

The temperature of the working fluid after the reheater is:

$$T_{eoo}(i) = \eta_{he} T_{eai}(i) + (1 - \eta_{he}) T_{eoi}(i) \tag{8}$$

where $T_{eoo}(i)$ is the temperature of the heat transfer fluid at the reheater outlet, in Kelvin (K).

Based on the energy balance equation, the heat absorbed by the air in the heater is:

$$Q_{he} = m_e c_a \left[T_{eao}(i) - T_{eai}(i)\right] \tag{9}$$

3.1.4. System efficiency. System efficiency refers to the efficiency of energy conversion, calculated using the formula:

$$\eta = \sum W_T / \sum W_C \tag{10}$$

$$W_T = m_{air,T}(h_{air,T,in} - h_{air,T,out})t_T \tag{11}$$

$$W_C = m_{air,C}(h_{air,T,out} - h_{air,T,in})t_C \tag{12}$$

Where η is the system efficiency; W is the expansion work or compression work; m_{air} is the Air mass flow rate; Subscripts: in = inlet, out = outlet; T = turbine, C = compressor; h = enthalpy.

3.2. Boundary condition

This article is based on the feasibility study design scheme of a developing 100 MW insulating compressed air energy storage power station currently, with the system boundary parameters of the power plant as illustrated in Table 1.

Table 1. System boundary parameters.

Serial number	Parameter	Value
1	Pressure loss of cooler, % (tube side/shell side)	0.59%/0.32%
2	Minimum temperature difference of cooler, °C	15
3	Cooling temperature of the cooler, °C	40
4	Cooling water temperature, °C	25
5	Cooling water pressure, °C	2
6	Isentropic efficiency of compressor, %	85/80/80/80
7	Energy storage duration, h	6
8	Compression series	4
9	Ambient temperature, °C	5
10	Environmental pressure, MPa	0.0683
11	Initial pressure of expansion section, MPa	7.62
12	Pressure loss of heater, % (tube side/shell side)	0.59,0.32
13	Minimum temperature difference of heater, °C	10
14	Isentropic efficiency of expander, %	88/88/88
15	Energy release duration, h	6
16	Expansion series	3

3.3. Establishment and verification of the thermodynamic model

3.3.1. Thermodynamic model. As shown in Figure 3, this article uses the Ebsilon software to establish a full-process model of a 100 MW power compression air energy storage system. On the basis of the geographical location and climate conditions of the project, the project design scheme includes four stages of compression (with the fourth stage compressor adopting variable frequency operation technology), three stages of expansion, an energy storage duration of 6 hours, and a reservoir storage pressure of 10 MPa. The project's storage method is underground hard rock cavern storage.

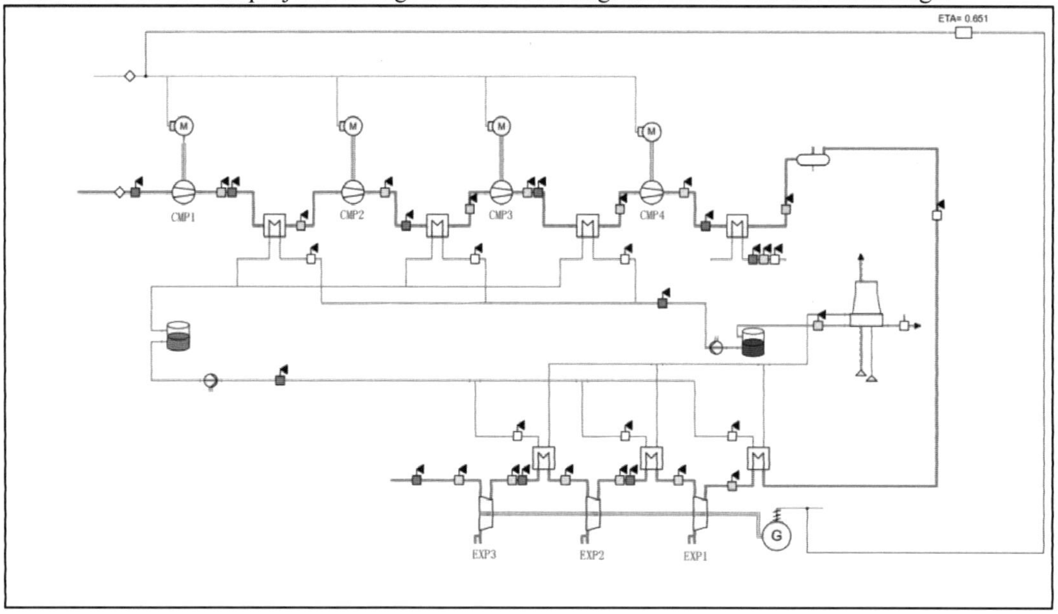

Figure 3. System thermodynamic model.

3.3.2. Model validation. A verification should be conducted to guarantee the accuracy of the established model. The models of key components (compressors, expanders, and heat exchangers) were validated by comparison. Using air as the working fluid, the input data for each component model comes from reference [11]. The calculated results can be referred to in Table 2.

Table 2. Model validation data comparison.

	parameter	Calculation data	Reference data	Error, %
Compressor	Inlet pressure, kPa	600	600	0
	Inlet temperature, K	298.15	298.15	0
	Outlet pressure, kPa	1920	1920	0
	Outlet temperature, K	436.75	437.16	-0.09
Expander	Inlet pressure, kPa	19661	19661	0
	Inlet temperature, K	437.16	437.16	0
	Outlet pressure, kPa	6144	6144	0
	Outlet temperature, K	328.57	328.96	-0.12
Heat Exchanger	Water side inlet temperature, K	298.15	298.15	0
	Gas side inlet temperature, K	436.75	437.16	-0.09
	Water side outlet temperature, K	431.75	432.16	-0.09
	Gas side outlet temperature, K	308.15	308.15	0

As shown in Table 2, the relative error between this result and that of the reference literature is small, demonstrating good consistency.

4. Simulation analysis of key parameters

4.1. Pressure ratio

Figure 4 illustrates the trend of system loop efficiency as a function of the pressure ratio. Within the study range, the cycle efficiency monotonically increases with the pressure ratio. Specifically, when the pressure ratio increases from 3.08 to 3.58, the cycle efficiency changes by 3.62%, rising from 60.44% to 64.06%.

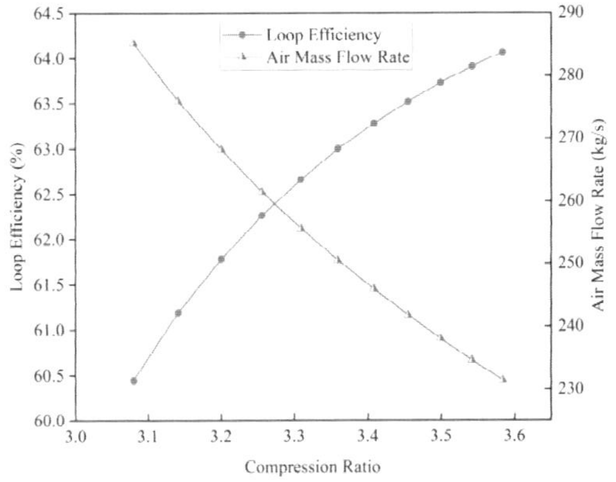

Figure 4. Variation Trend of loop efficiency and air mass flow with pressure ratio.

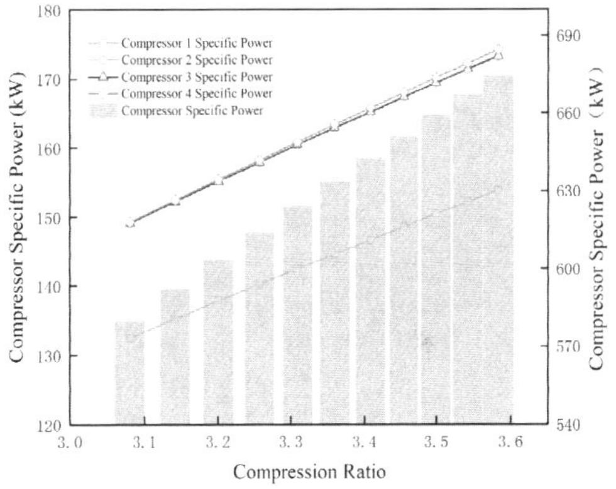

Figure 5. Variation Trend of specific power of compressors at all levels and specific power during compression with pressure ratio.

With the changing of pressure proportion, the mechanism of energy conversion in the system is as follows: Figure 4 illustrates the trends of specific power and specific compression work for each stage of the compressor as functions of the pressure proportion. As the pressure ratio increases, the range of pressure changes in the air across each stage of the compressor also increases, which results in the particular power consumption of each stage of the compressor following the increasing trend shown in Figure 5. The particular compression work rises with the pressure ratio, leading to a decreasing trend in cycle efficiency. On the other hand, the particular expansion work and the pressure ratio increase simultaneously. Under a constant power output of 100 MW, the mass flux of air in the system reduces while the pressure ratio drops conversely, as shown in Figure 5, which results in an increasing trend in cycle efficiency. Primarily influenced by the decrease in mass flow rate with the rise in pressure ratio, the system cycle efficiency exhibits the increasing trend shown in Figure 4.

4.2. Expansion ratio

Figure 6 illustrates the trend of loop efficiency as a function of the expansion ratio. Within the study range, the cycle efficiency monotonically increases as the expansion ratio increases. Specifically, when the expansion ratio increases by 0.624, the cycle efficiency increases by 4.35%, from 60.40% to 64.75%.

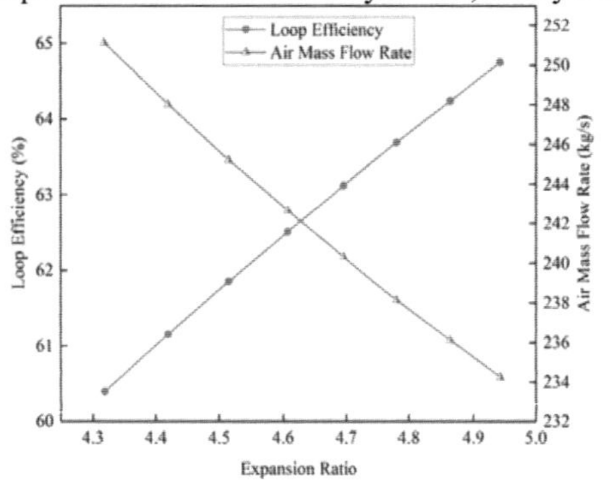

Figure 6. Variation Trend of loop efficiency and air mass flow with expansion ratio.

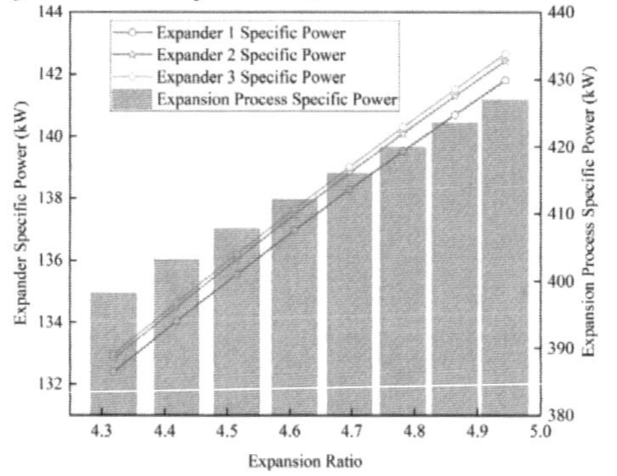

Figure 7. Variation Trend of the specific power of expanders at all levels and specific power during expansion with expansion ratio.

As the expansion ratio varies, the structure for energy transformation in this system is as follows: Figure 7 shows the trends of specific power and specific expansion work for each stage of the expander as functions related to the expansion ratio. As the expansion ratio increases, the range of pressure changes in the air across each stage of the expander also increases, which results in the specific power and specific expansion work for each stage of the expander following the increasing trend shown in Figure 7. Under a constant system power output, the air mass flux in this system declines, as shown in Figure 6, dropping from 251.13 kg/s to 234.24 kg/s, while the specific compression work does not change with the expansion ratio. Therefore, primarily influenced by the decrease in mass flux, the cycle efficiency exhibits the increasing trend shown in Figure 6.

4.3. Heat storage temperature

Figure 8 illustrates the trend of loop efficiency with changes in heat storage temperature. As the heat storage temperature is enhanced, the cycle efficiency first rises and then falls. Specifically, when the heat storage temperature ascends from 215.26 °C to 222.32 °C, the system cycle efficiency increases from 67.03% to 67.61%. However, as the heat storage temp keeps rising, the cycle efficiency starts to drop.

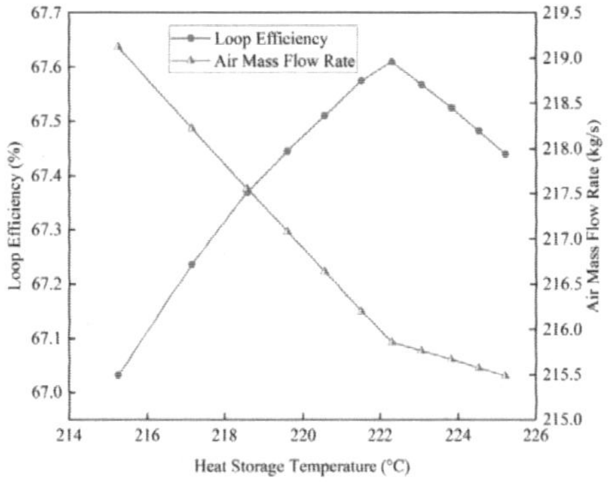

Figure 8. Variation trend of loop efficiency and air mass flow with heat storage temperature.

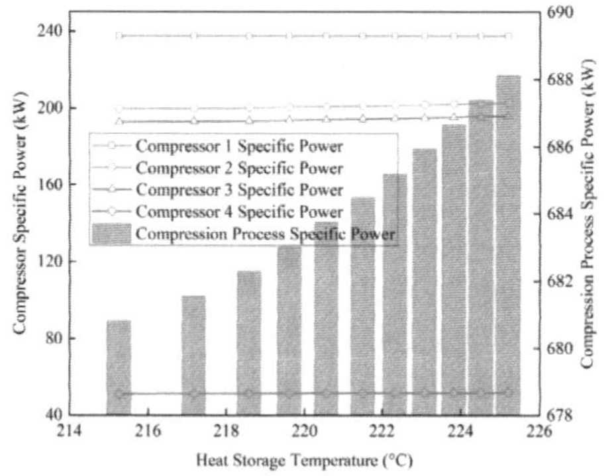

Figure 9. Variation trend of specific power of compressors at all levels and specific power during compression with stored heat temperature.

Journal of Physics: Conference Series **3031** (2025) 012003 doi:10.1088/1742-6596/3031/1/012003

With the variations in stored heat temperature, the energy conversion mechanism in the system is as follows: A side-by-side analysis is conducted to illustrate the tendency of the specific power at each stage of the compressor and the specific power in the expansion process, as shown in Figure 9. With the boost of the heat storage temperature, the specific power at each stage of the compressor increases, resulting in an increase in compression power consumption and a decrease in cycle efficiency. Moreover, raising the stored heat temperature can enhance the specific power output of the expansion process, and the air mass flow shows a downward trend in Figure 8, resulting in an increase in the system cycle efficiency. When the stored heat temperature increases from 215.26 °C to 222.32 °C, the system efficiency will continue to increase due to the main influence of the latter factor. When the stored heat temperature further increases, the cycle efficiency will decrease according to the primary effect of the former factor.

4.4. Heat exchange temperature difference

The ways of cycle efficiency changing with heat transfer temperature difference are demonstrated in Figure 10. An increase in heat transfer temperature difference leads to a monotonic decrease in system cycle efficiency. Specifically, when the heat transfer temperature difference jumps from 7 °C to 15 °C, cycle efficiency diminishes by 2.35%, decreasing from 68.66% to 66.31%.

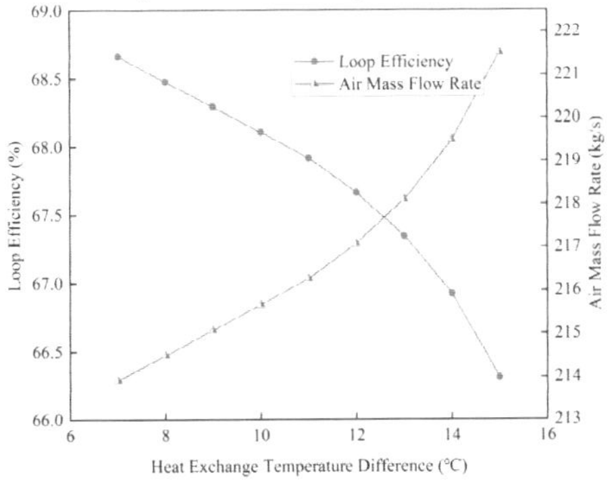

Figure 10. Variation trend of loop efficiency and air mass flow with heat exchange temperature difference.

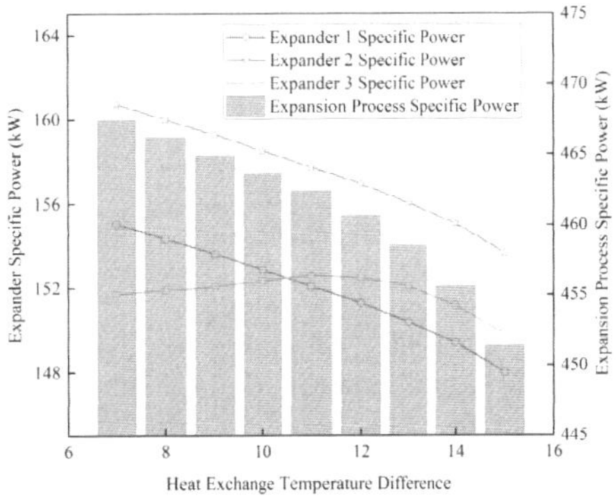

Figure 11. Variation trend of specific power and expansion process specific power of expanders at all levels with heat exchange temperature difference.

As the heat exchange temperature difference varies, the mechanism of energy conversion within the system is as follows: Figure 11 illustrates the trends of specific power and specific expansion work for each stage of the expander as functions of temperature differential in heat transfer. As the heat transfer temperature gradient increases, the system's heat storage temperature decreases, causing the specific expansion power output to exhibit a decreasing trend. Under a constant system power output of 100 MW, the air mass flow rate shows an increasing trend, as depicted in Figure 10. Additionally, fluctuations in heat transfer thermal difference do not affect the specific compression work. Therefore, under the crucial impact of the increasing air mass flow rate, the system cycle efficiency presents a decreasing trend, as shown in Figure 10.

5. Conclusion

This article builds up a 100 MW advanced adiabatic CAES system utilizing Ebsilon software and conducts an in-depth analysis of the energy conversion mechanisms and their impact on system efficiency during the variation of key parameters such as pressure ratio, expansion ratio, storage temperature, and heat transfer temperature difference. The conclusions are as follows:

1) Increasing the compression ratio has opposite effects on system compression power consumption and working fluid flow rate. Through comprehensive analysis, the dominant effect is a decrease in mass

Journal of Physics: Conference Series **3031** (2025) 012003 doi:10.1088/1742-6596/3031/1/012003

flow rate with increasing compression ratio, while there will be a boost in the overall system efficiency as the compression ratio rises.

2) A jump in the expansion ratio causes a drop in air mass within the system under a constant 100 MW power output (from 251.13 kg/s to 234.24 kg/s), and the cycle efficiency shows a rising trend.

3) As the storage temperature jumps, the cycle efficiency of this system initially increases and it falls after peaking; the critical temperature fluctuates around 222.3 °C.

4) The system's storage temperature will descend after the climb of the heat exchange temperature difference. Under a system power output of 100 MW, the air mass flow rate goes up; therefore, the system cycle efficiency decreases according to the dominant effect of increasing air mass flow rate.

Acknowledgement

Funding: The research was funded by the National Natural Science Foundation of China (Grant No. U23B20145) and the Key Research Project of Powerchina (DJ-HXGG-2023-09).

References

[1] Liu, X.C.; Mei, S.W.; Ding, R.C.; Zhong, S.Y.; Zhang, J.F.; Xie, N.N.; Chang, Y.; Zhang, T. Present situation, Development trend and Application Prospect of compressed Air Energy Storage Engineering. Electr. Power Autom. Equip. 2023, 43, 38–47+102.

[2] Guo H, Kang H, Xu Y, et al. Review of Coupling Methods of Compressed Air Energy Storage Systems and Renewable Energy Resources[J]. Energies, 2023,16(12):

[3] ROOS P, HASELBACHER A. Analytical modeling of advanced adiabatic compressed air energy storage: Literature review and new models[J]. Renewable and Sustainable Energy Reviews, 2022, 163: 112464.

[4] SONG J, PENG X, FANG X, et al. Thermodynamic analysis and algorithm optimisation of a multi-stage compression adiabatic compressed air energy storage system[J]. Thermal Science and Engineering Progress, 2020,19:100598.

[5] Dupin V, Teixeira D. Advanced adiabatic compressed air energy storage systems dynamic modelling: Impact of the heat storage device [J]. Heliyon,2025,11(1): e40730-e40730.

[6] Yucheng F. The feasibility of conducting compressed air energy storage system [J]. E3S Web of Conferences,2025,606.

[7] Lujing H, Huan G, Yujie X, et al.Influence of design point on off-design and cycling performance of compressed air energy storage systems-from key processes to the whole system [J]. Journal of Energy Storage, 2023, 57.

[8] Zhang Xinjing, Gao Ziyu, Zhou Bingqian, et al. Advanced compressed air energy storage systems: fundamentals and applications[J]. Engineering, 2024.

[9] Cui S, Chen L, Chen S, et al. Dynamic modeling and analysis of compressed air energy storage for multi-scenario regulation requirements[J]. Journal of Energy Storage,2024,100(PA):113227-113227.

[10] Li F, Shu Y, Liu X, et al. Study on the influence of expansion section parameters on the performance of compressed air energy storage system [J]. Journal of Physics: Conference Series, 2024, 2703(1):

[11] Rehman S, Al-Hadhrami M L, Alam M M. Pumped hydro energy storage system: A technological review[J]. Renewable and Sustainable Energy Reviews, 2015, 44586-598.

Journal of Physics: Conference Series **3031** (2025) 012004 doi:10.1088/1742-6596/3031/1/012004

Thermodynamic simulation and economic performance of liquid carbon dioxide energy storage coupled with a low-temperature molten salt heat storage system

Qiwen Sun, Jinchen Cao, Jingshun Zhang, Xing Zheng, Xiaojian Wang*

Department of Energy and Power Engineering, College of Electrical Engineering, Xinjiang University, Urumqi 830047, China

*Corresponding author's e-mail: xjwang@xju.edu.cn

Abstract. To address environmental challenges and grid stability issues caused by high renewable energy penetration, liquid carbon dioxide energy storage (LCES) has emerged as a promising solution due to its high energy density and geographical flexibility. However, Water-based thermal storage in LCES systems suffered significant heat loss from temperature limitations. This study proposes an innovative LCES system integrated with low-temperature molten salt thermal storage (LTMS-LCES), which utilizes surplus wind power to drive compressors during off-peak periods, storing both electrical energy and compression heat in molten salt. The analysis reveals that increasing the pressure of the high-pressure tank enhances both energy storage density (EVR) and heat storage efficiency (HSE), but it adversely affects the round-trip efficiency (RTE) and the levelized cost of electricity (LCOE). Furthermore, compared to the conventional LCES system, the proposed LTMS-LCEC system achieves a 15% improvement in RTE and an EVR of 11.2 kWh/m^3. This work provides critical insights for commercializing molten salt-enhanced LCES technology.

1. Introduction

The rapid pace of social development has significantly increased electricity demand, driving up fossil fuel consumption, which in turn amplifies carbon emissions and intensifies environmental pressures[1,2]. This has prompted governments to prioritize carbon emission reductions and accelerate the adoption of renewable energy[3]. However, the instability of renewable energy sources has become a barrier when they are integrated into the grid on a large scale. As a result, this has led to the development of large-scale energy storage systems (L-ESS), which hold great promise and align with the requirements of green development[4].

Compared to large-scale ESS technologies like pumped hydro and compressed air energy storage (CAES), compressed carbon dioxide energy storage (CCES) has gained attention due to its favorable critical point (304.12 K, 7.38 MPa) and advantageous physical properties (low viscosity, higher density) in the supercritical state[5]. An increasing number of researchers are focusing on liquid carbon dioxide energy storage (LCES) technology, as it can effectively enhance the energy density of the system[6]. Sun et al. [7]reported that LCES can achieve an RTE of 51.45% and an EVR of 22.21 kWh/m^3 under typical conditions.

However, most existing LCES systems use water as the medium to store the heat generated during the compression process. While water, being a low-cost and easily accessible working fluid, offers significant value, its heat absorption capacity diminishes beyond 100°C due to the limitations imposed by its critical temperature. Therefore, it is essential to explore alternative heat storage media to replace water.

This paper presents a low-temperature molten salt heat storage system integrated with LCES. By developing a mathematical model of the entire system, the impact of key parameters on performance

Journal of Physics: Conference Series **3031** (2025) 012004 doi:10.1088/1742-6596/3031/1/012004

and energy loss distribution is analyzed. An economic analysis, based on valley-hour electricity pricing, is also conducted. This work provides insights into the industrialization of LCES systems coupled with low-temperature molten salt heat storage.

2. Low-temperature molten salt heat storage coupled liquid carbon dioxide energy storage system (LTMS-LCES)

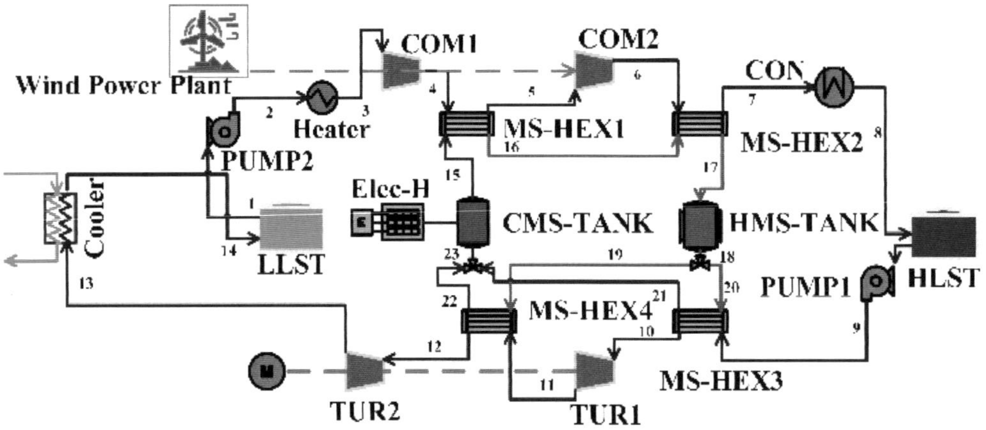

Figure 1. Technical schematic of the novel LTMS-LSES design framework.

2.1 System description

The detailed LTMS-LCES system process is depicted in Figure 1. The proposed system is composed of LCES energy storage and LTMS heat storage subsystems. Specific comprehensive system working processes are as follows:

The LTMS-LCES system consists of two main stages: the charging process and the discharging process. During off-peak hours, low-pressure liquid CO_2 is retrieved from a low-pressure storage tank (LLST) and heated to convert it into a saturated gas at the same pressure. Two-step pressurization incorporates inter-stage thermal management, with compression-derived heat flux being captured through cryogenic salt-based cooling units (MS-HEX1-2). The cooled gas then enters a condenser (CON), where it is liquefied and subsequently stored in the high-pressure liquid storage tank (HLST).

During the peak hours, CO_2 stored in the HLST in liquid form passes through PUMP1 and enters the turbine (TUR) to work. The adjusted CO_2 then enters MS-HEX3-4, where it absorbs heat from the molten salt in the HMS-TANK and flows into TUR1-2 to expand and generate electricity. After performing work, the CO_2 is cooled in the cooler to a low-pressure liquid state and stored in the LLST.

2.2 Mathematical model

In this paper, a comprehensive analysis is carried out based on the equilibrium equation of each component.

$$\sum \dot{E}_Q - \sum \dot{E}_{\dot{w}} = \sum \dot{E}_{out} - \sum \dot{E}_{\dot{w}} + \dot{E}_D \tag{1}$$

where $\dot{E}_Q$ and $\dot{E}_W$ represent the exergy generated by heat exchange and work, respectively. $\dot{E}_D$ means the exergy destruction rate. And all exergy flow considers the time variation.

2.3 Economic model

Cost-benefit evaluation provides a validated methodology for assessing system development feasibility in attaining operational engineering implementation. Based on the Chemical Engineering Plant Cost Index (CEPCI), the cost of each component is realized by setting a base year, taking into account the impact of the time factor. The specific formula is provided, where CEPCI2022 = 816.

$$Z_k = Z_{\text{original}} \cdot (\text{CEPCI}_{2022} / \text{CEPCI}_{\text{original}}) \quad (2)$$

$$Z_{equip,tot} = \sum_{k=1}^{n} Z_k \quad (3)$$

2.4 System performance criteria

This study employs round-trip efficiency (RTE) and energy released per unit volume of energy storage (EVR) as primary metrics to quantify the operational efficiency of the developed energy storage system. RTE correlates charge-phase power consumption with discharge-phase usable energy output, and the formula is expressed as follows:

$$RTE = \frac{\int_0^{t_{\text{discharge}}} \dot{W}_{net}\, dt}{\int_0^{t_{\text{char}}} \dot{W}_{input}\, dt} \quad (4)$$

EVR is defined as the ratio of energy released during the discharge process to the total storage volume of the system. Its definition is as follows:

$$EVR = \frac{\int_0^{t_{\text{discharge}}} \dot{W}_{\text{net}}\, \text{dt}}{V_{\text{tot}}} \quad (5)$$

The heat storage efficiency (HSE) is defined as the ratio of the actual heat absorbed (Q_{ah}) by the molten salt to the theoretical heat absorbed (Q_{th}) by the molten salt, and the equation is as follows:

$$HSE = \frac{\sum_{i=1}^{n} Q_{ah}(i)}{\sum_{i=1}^{n} Q_{th}(i)} \quad (6)$$

For the economic analysis, the levelized cost of electricity (LCOE) is selected, expressed as.

$$LCOE = \frac{Z_{equip,tot} \cdot \text{CRF} + Z_{\text{OM}} + Z_{\text{ele}} + Z_{LC} + Z_{CF}}{\dot{W}_{\text{net}} \cdot \tau} \quad (7)$$

wherein τ denotes annualized operational duration while CRF corresponds to the capital recuperation coefficient.

3. Results and discussion

3.1 Basic design performance

In this part, the LTMS-LCES system is simulated under the design condition. All the working fluid data in the system comes from the REFPROP software. The simulation of the whole system is completed in MATLAB. All initial parameters of the system input and low-temperature molten salt parameters are presented in Table 1.

Table 1. Typical operating condition of the proposed LTMS-LCES system.

Items	Parameters	Value	Parameters	Value
LCES	Ambient temperature (K)	298.15	Isentropic efficiency of the turbine (%)	88
	Ambient pressure (MPa)	0.101	Isentropic efficiency of the compressor (%)	85
	Pressure of HLST (MPa)	15	Isentropic efficiency of the pump (%)	75
	Temperature of HLST (K)	293.15	Charge time (h)	6
	ΔP_{pump} (MPa)	2	Discharge time (h)	6
	Valley electricity prices (kW·h/$)	0.052	System operating life	350
Molten salt	Composition ($NaNO_3/KNO_3/NaNO_2$)	(7/45/48)	Preliminary temperature (K)	413.15
	Density ($g \cdot cm^{-3}$)	1.9	Specific heat capacity ($kJ \cdot (kg \cdot K^{-1})$)	1.45

3.2 Parametric analysis

Two key parameters, storage pressure of HLST (P_{HLST}) and inlet temperature of molten salt (T_{msin}), are selected to reveal their effects on system performance. These two parameters have significant effects on energy storage and the efficiency of the system. While analyzing another parameter, the other parameter remains the same as the design condition.

3.2.1 Effects of the P_{HLST}.

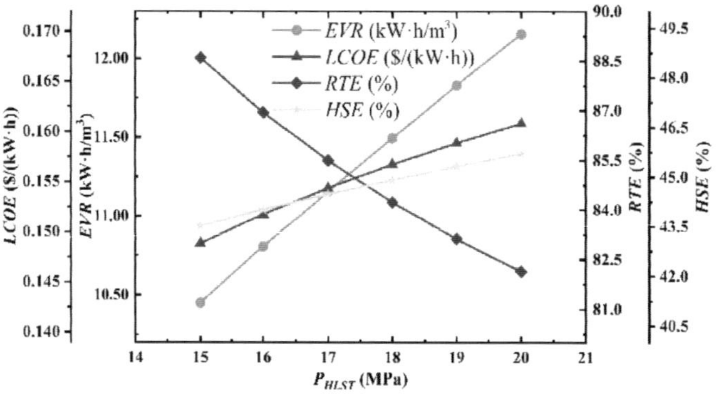

Figure 2. Influence of the pressure of HLST (P_{HLST}) on system performance.

Figure 2 reveals that HSE, EVR, and LCOE all increase while RTE decreases in the process of P_{HLST} increase, indicating that high P_{HLST} will greatly affect system efficiency. During the charging phase, the rise in P_{HLST} means that the pump pressure rises and the compressor outlet pressure is increased, which will inevitably increase the overall power consumption of the system. Besides, with the compressor pressure increasing, the electric heater consumption in the molten salt heat absorption process is greater, leading LCOE to 0.161 ($/kW·h). This is the main reason for the 5% increase in HSE. However, because the pressure of HLST increases, the density of supercritical CO_2 increases, and the mass of CO_2 in the system is constant, it will lead to the decrease of V_{HLST}, and finally, EVR presents an upward trend.

3.2.2 Effects of the temperature of molten salt inlet (T_{msin}). Figure 3 demonstrates how T_{msin} influences the performance of the system. As Tmsin increases, the fixed pressure ratio and constant heat release during compression lead to minimal variations in EVR and LCOE, given the unchanged CO_2 inlet state. However, the heat exchanger temperature fails to match the CO_2 outlet temperature optimally, reducing heat storage efficiency (HSE). Although electric heater power consumption rises with Tmsin, its cost impact on LCOE remains minor. The inefficient thermal energy recovery in compression processes stems from significant thermal gradients between opposing fluid pathways. Concurrently, the limited efficiency of heat exchangers eventually leads to the reduction of the final temperature of CO_2. Consequently, the RTE is reduced.

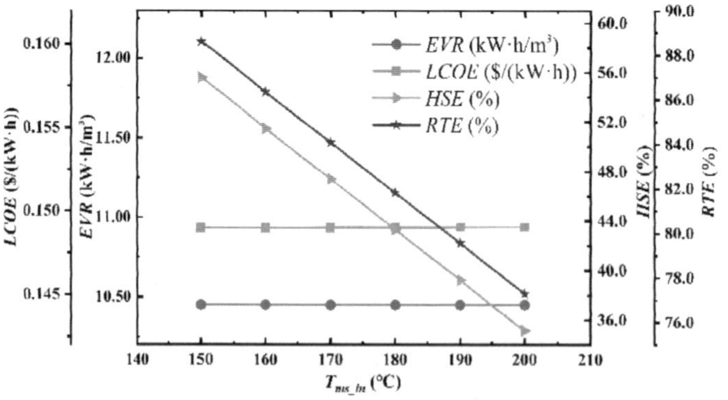

Figure 3. Effects of the molten salt inlet temperature (T_{msin}) on system performance.

3.3 Economic analysis

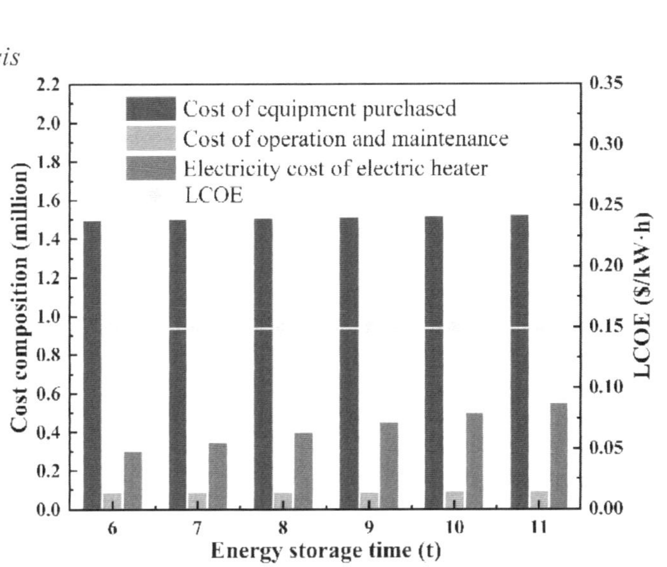

Figure 4. Effects of the energy storage time on system cost and economic performance.

Figure 4 illustrates the impact of storage time on the system costs and overall economic performance (LCOE). During the charging process, an increase in energy storage time extends the duration required for the electric heater to heat the molten salt to the dissolved state, leading to higher electricity costs for the electric heater. Although the operation and maintenance costs rise by 7.3%, the change is not significant. The equipment purchase cost remains largely unaffected by charging time, as the system structure stays the same. In the LCOE calculation, since the electricity consumption of the electric heater is sourced from low valley prices, it represents a small portion of the total cost. Consequently, storage time has minimal impact on LCOE.

3.4 Exergy analysis

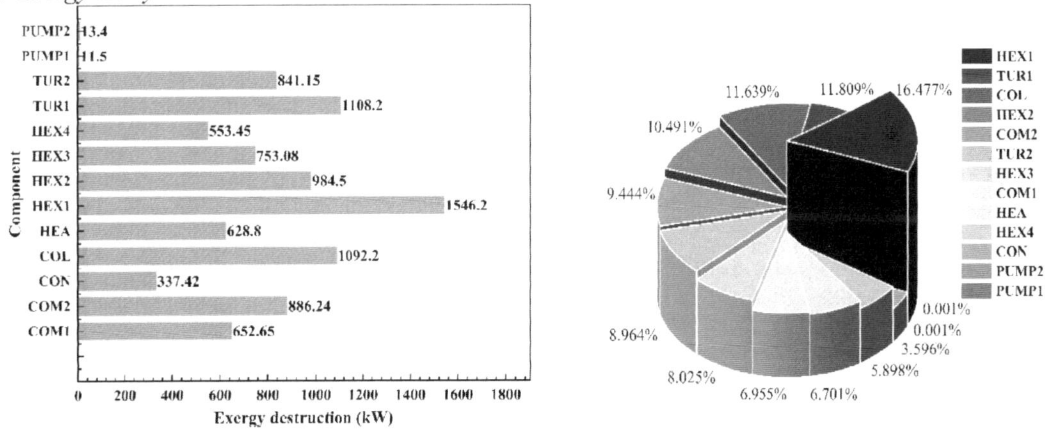

Figure 5. Exergy destruction of each component of the proposed LTMS-LCES system.

The exergy loss and its proportion for each component of the system are calculated based on the basic operating parameters, as shown in Figures 5a and b. The total exergy destruction of the LTMS-LCES system is 9.41 MW. HEX1 exhibits the highest exergy loss at 1.55 MW, accounting for 16.48% of the total. These rankings are primarily due to the large temperature difference between the initial molten salt inlet temperature and the CO_2 outlet temperature from COM1, which leads to significant heat transfer losses in the heat exchanger and limits its efficiency. Furthermore, the higher the outlet temperature of TUR1, the greater the inlet temperature at the hot end of HEX4, which increases the heat transfer temperature difference in COL. Although COL and HEX efficiencies are similar, COL experiences a large exergy loss due to these factors. In contrast, the exergy losses for PUMP1 and PUMP2 are minimal and negligible.

4. Conclusions

Integrating low-temperature molten salt thermal storage into LCES systems, this research establishes an innovative LTMS-LCES configuration, with particular emphasis on evaluating its energy conversion efficiency and cost-effectiveness metrics. The conclusions are summarized as follows:

(1) An increase in P_{HLST} is beneficial to EVR and HSE by 16.3% and 5%, respectively, but negatively affects RTE and LCOE by 82.14% and 0.161$/kWh, respectively.

(2) Exergy analysis results show that HEX1 has the highest loss, and appropriate T_{msin} matching the CO_2 inlet temperature should be considered to prevent excessive heat transfer temperature difference.

(3) The system has demonstrated competitive performance, outperforming other reported CCES systems in several respects with the adoption of the LTMS system.

Acknowledgments

This work is sponsored by the Natural Science Foundation of Xinjiang Uygur Autonomous Region (No. 2022D01C420, No.2023D01C197). Major Science and Technology Projects of Xinjiang Uygur Autonomous Region (No. 2024A01004-1), Tianchi Talents Introduction Plan of Xinjiang Uygur Autonomous Region, State Key Laboratory of Clean Energy Utilization Open Fund Project (No. ZJUCEU2023020), and Guangdong Provincial University Innovation Team Project (No. 2023KCXTD038).

References

[1] Calcaterra M, Aleluia Reis L, and Fragkos P, et al. (2024) Reducing the cost of capital to finance the energy transition in developing countries. Nat Energy, 9:1241–51.

[2] Danish, Baloch MA, Mahmood N, Zhang JW. (2019) Effect of natural resources, renewable energy, and economic development on CO2 emissions in BRICS countries. Sci Total Environ, 678:632–8.

[3] Cherp A, Vinichenko V, and Tosun J, et al. (2021) National growth dynamics of wind and solar power compared to the growth required for global climate targets. Nat Energy, 6:742–54.

[4] Shamsi SSM, Barberis S, and Maccarini S, et al. (2024) Large-scale energy storage systems based on carbon dioxide thermal cycles: A critical review. Renew Sustain Energy Rev, 192:114245.

[5] Liu H, He Q, and Borgia A, et al. (2016) Thermodynamic analysis of a compressed carbon dioxide energy storage system using two saline aquifers at different depths as storage reservoirs. Energy Convers Manag., 127:149–59.

[6] Wang M, Zhao P, and Yang Y, et, al. (2015) Performance analysis of energy storage system based on liquid carbon dioxide with different configurations. Energy., 93:1931–42.

[7] Sun W, Liu X, and Yang X, et al. (2021) Design and thermodynamic performance analysis of a new liquid carbon dioxide energy storage system with low-pressure stores. Energy Convers Manag., 239:114227.

Journal of Physics: Conference Series **3031** (2025) 012005 doi:10.1088/1742-6596/3031/1/012005

Economic and resilient operation strategy of residential area based on battery swapping station

Qingkai Sun*, Zeyu Ding, Menghua Fan, Zheng Zhao, Qiuyang Ma, Chen Lv

State Grid Energy Research Institute Co., Ltd, Beijing, China

*Corresponding author's email: sunqingkai_123@163.com

Abstract. With the rapid proliferation of electric vehicles (EVs), the EV population in residential areas and the number of associated battery swapping stations (BSS) have surged. Uncoordinated charging practices can intensify peak-valley load disparities, thereby undermining grid stability and escalating electricity costs for consumers. While EVs and BSSs are equipped with vehicle-to-grid (V2G) capabilities that enable grid support during emergencies and enhance distribution network resilience, optimizing their charging and discharging strategies across diverse grid conditions remains a critical challenge. In this study, we identify three distinct grid states with corresponding optimization periods, develop bespoke optimization models and operational strategies for each state, and validate the proposed approach through case studies that illustrate improved coordination, mitigated load fluctuations, and enhanced grid resilience.

1. Introduction

The use of EVs can reduce the consumption of fossil fuels, reduce CO_2 emissions, and reduce the cost of vehicle use [1]. To cope with the energy crisis and reduce the carbon emissions of the transportation industry, countries around the world are vigorously promoting the use of EVs, hoping to achieve low-carbon development of the transportation industry through electrification [2]. In recent years, driven by the policies of various countries, the world's EV industry has flourished; not only has the number of EVs increased rapidly, but also the number of BSSs has increased significantly [3]. The charging and swapping behavior of EVs and BSSs has significant uncertainty in time and space. Charging and swapping batteries of a large number of EVs during peak hours will cause a load surge, resulting in energy supply and demand imbalance, relay protection, and other problems [4]. However, the charging load of EVs and BSSs has significant flexibility and adjustability. If the charging and swapping behavior of EVs is guided by reasonable means. The situation will be greatly improved by making EVs charge and swap batteries at the peak period of new energy output or the valley period of power load, which can promote the absorption of new energy and stabilize the power grid load [5].

In addition, EVs and BSSs also have energy storage characteristics, and with the help of the V2G function of EVs, they can achieve discharge to the grid. When the power system has an insufficient power supply, such as insufficient new energy output or emergencies such as power outage, EVs and BSSs can ensure the safe operation of the power grid and improve the resilience of the distribution network through discharging to the power grid [6].

There is a lot of literature on the impact of charging and discharging of EVs in residential areas on the economy of the distribution network and the optimal operation of BSS. Based on EV charging load demand prediction, Chen et al. [7] propose an optimization method for power grid scheduling. Hou et al. [8] propose a long-time scale charging and discharging scheduling strategy for electric vehicles under the combined demand response of incentive and price, which realizes the full mobilization of the charging and discharging potential of electric vehicles. Wang et al. [9] studies the orderly control strategy of EV charging load based on the impact of pricing model on the charging strategy of power grid orderly operation, which aims to prevent the overload of distribution transformers and maximize the charging income.

In Sun et al.'s study [10], the charging scheduling problem is represented as a constrained Markov decision process (MDP), and a strategy to minimize the charging cost of a BSS is obtained by using the standard Lagrange algorithm and the dynamic programming algorithm. Wu et al. [11] propose a virtual queue of battery demand and develop a dynamic charging scheduling algorithm based on real-time demand by using a Lyapunov optimization framework to minimize the backlog of battery demand queue. Ko et al. [12] propose a scheduling algorithm for EV-assisted BSS to provide battery swapping service. BSS swaps batteries in consideration of the distribution of electricity price and EV arrival rate.

However, most of the existing studies focus on the impact of EVs or BSS in residential areas on the power grid alone, without considering the mutual impact of the two. At the same time, it rarely considers the different needs of residential power grids for EVs and BSSs under different grid states. Therefore, the paper first divides the state of the power grid into three types according to the demand under different conditions, then establishes multi-stage optimization models of different distribution network states, and proposes the economic and resilient operation strategy process of residential areas based on EV and BSS. Finally, the paper tests the proposed model and operation strategy by case study, and verifies its effectiveness.

2. State classification of distribution network in residential area

According to the different demands of residential users and power systems for power supply, the state of the distribution network is divided into three types: economic dispatch, proactive defense, and load restoration.

Most of the time, the distribution network works normally and is in the state of economic dispatch. Users in residential areas expect to reduce the cost of electricity as much as possible while fully meeting their electricity demand, while the power supply side hopes to reduce the peak-valley difference of electricity load to ensure the safety and reliability of the power supply. When predictable natural disasters such as typhoons and snowstorms or planned power outages are about to occur, users in residential areas first hope to fully charge the energy storage devices as much as possible to cope with the coming power outages. At this time, the distribution network is in the state of proactive defense. When the entire residential station area is in a power outage, to maintain the continued operation of some important loads, EVs and BSSs in the residential area discharge through the V2G function. And the distribution network is in the state of load restoration.

To ensure the real-time performance of the optimization strategy, it is necessary to update the relevant data for re-optimization after a certain time interval to achieve the purpose of rolling optimization. After determining the state of the power grid in the current period, it is necessary to first determine the schedule horizon. Different schedule horizons under the distribution network states are selected as follows:

1) Economic dispatch: The economic dispatch state is the state that the distribution network is in during its normal working time, so the schedule horizon in this state should be consistent, no matter what the current time is. Considering the accuracy of the prediction of data in the optimization model, the schedule horizon is generally fixed as 24 hours, and the time step is generally selected as 15 minutes. Therefore, the schedule horizon is usually fixed as 96 time steps in the economic dispatch state.

2) Proactive defense: The proactive defense state is the intermediate state between economic dispatch and load restoration. Therefore, even if the power company announces the time of planned power outages many days in advance, residential areas generally only need to take proactive defensive responses within one day before the power outage. Considering the short duration of the entire proactive defense state, the schedule horizon is set to the time between the current time and the start time of the power outage. If the power outage time is fixed, the schedule horizon will gradually decrease with time, which represents the gradual arrival of power outage events.

3) Load restoration: In this state, the main network no longer provides electricity to the residential area. To keep the important load working, the residential area can meet its electricity demand as much as possible through the energy storage of the batteries in EVs and BSS. Considering that the outage time

Journal of Physics: Conference Series **3031** (2025) 012005 doi:10.1088/1742-6596/3031/1/012005

is generally short, the schedule horizon should be set to the time between the current time and the end time of the outage.

3. Optimization models for residential area in different distribution network states

3.1 Economic dispatch model

Due to the limited space in the paper and the economic dispatch of EV and BSS in residential areas has been studied in another paper [13]. Therefore, the optimization model of economic dispatch state is not introduced in detail in this paper, which can refer to Li et al.'s study [13] and the models of other states in the following paper.

3.2 Proactive defense model

3.2.1 First-stage optimization model.

$$\max \sum_{t \in S_{t_c}(x)} P_{\text{buy}}^{x}(t)\Delta t \tag{1}$$

Where $S_{t_c}(x)$ represents the set of discrete time intervals starting from t_c in schedule horizon x. $P_{\text{buy}}^{x}(t)$ is the total purchased electrical power in time interval t of schedule horizon x. The objective function of the first stage is to maximize the sum of the purchased electricity in the residential area during the scheduling horizon. Considering that the load of the residential area and the power demand of BSS will not vary with the change of the distribution network state, the setting of the objective function will make the battery of EVs and BSS in the residential area charge as much as possible, to cope with the impending power outage.

$$u_{\text{BSS,b}}^{\text{swap},x}(i,t) \le Soc_{\text{BSS,b}}^{x}(i,t) - Soc_{\max} + 1, i \in N_{\text{BSS,b}}(x), t \in S_{t_c}(x) \tag{2}$$

$$\sum_{j \in N_{\text{BSS,ev}}(x)} u_{\text{BSS}}^{\text{swap},x}(i,j,t) = u_{\text{BSS,b}}^{\text{swap},x}(i,t), i \in N_{\text{BSS,b}}(x), t \in S_{t_c}(x) \tag{3}$$

$$\sum_{i \in N_{\text{BSS,b}}(x)} u_{\text{BSS}}^{\text{swap},x}(i,j,t) = u_{\text{BSS,ev}}^{\text{swap},x}(j,t), j \in N_{\text{BSS,ev}}(x), t \in S_{t_c}(x) \tag{4}$$

$$\begin{gathered} Soc_{\text{BSS}}^{\text{swap},x}(i,j,t) = u_{\text{BSS}}^{\text{swap},x}(i,j,t)(Soc_{\max} - Soc_{\text{BSS,ev}}^{\text{swap},x}(j)) \\ i \in N_{\text{BSS,b}}(x), j \in N_{\text{BSS,ev}}(x), t \in S_{t_c}(x) \end{gathered} \tag{5}$$

$$\sum_{j \in N_{\text{BSS,ev}}(x)} Soc_{\text{BSS}}^{\text{swap},x}(i,j,t) = Soc_{\text{BSS,b}}^{\text{swap},x}(i,t), i \in N_{\text{BSS,b}}(x), t \in S_{t_c}(x) \tag{6}$$

$$Soc_{\min} \le Soc_{\text{BSS,b}}^{x}(i,t) \le Soc_{\max}, i \in N_{\text{BSS,b}}(x), t \in S_{t_c}(x) \tag{7}$$

$$\begin{gathered} Soc_{\text{BSS,b}}^{x}(i,t+1) = Soc_{\text{BSS,b}}^{x}(i,t) + P_{\text{BSS,b}}^{\text{ch},x}(i,t)\eta^{\text{ch}}\Delta t / E_{\text{BSS,b}}^{\text{cap}} \\ i \in N_{\text{BSS,b}}(x), t \in S_{t_c}(x) \end{gathered} \tag{8}$$

$$0 \le P_{\text{BSS,b}}^{\text{ch},x}(i,t) \le u_{\text{BSS,b}}^{\text{ch},x}(i,t) P_{\text{BSS,b}}^{\max,\text{ch}}, i \in N_{\text{BSS,b}}(x), t \in S_{t_c}(x) \tag{9}$$

$$\sum_{i \in N_{\text{BSS,b}}(x)} P_{\text{BSS,b}}^{\text{ch},x}(i,t) \le P_{\text{BSS,b}}^{\max,\text{ch,sum}}, t \in S_{t_c}(x) \tag{10}$$

$$u_{\text{BSS,b}}^{\text{ch},x}(i,t) + u_{\text{BSS,b}}^{\text{swap},x}(i,t) \le 1, i \in N_{\text{BSS,b}}(x), t \in S_{t_c}(x) \tag{11}$$

Equations (2)-(11) refer to the charge and swapping constraints of BSS, where $N_{\text{BSS,b}}(x)$ and $N_{\text{BSS,ev}}(x)$ are sets of batteries of EVs with swapping function and BSS, respectively. $u_{\text{BSS,b}}^{\text{swap},x}(i,t)$ and $u_{\text{BSS,b}}^{\text{ch},x}(i,t)$ are 0/1 variables representing whether the battery i of the BSS charges and swaps with the battery of the EV. Equation (11) represents that a battery of BSS cannot charge and swap simultaneously.

$Soc^{x}_{\text{BSS,b}}(i,t)$ is the state of charge (SOC) of battery i in BSS, and $Soc_{\max}$ / $Soc_{\min}$ or is the maximal/minimal SOC of all batteries, so (2) represents that a battery cannot be swapped if it is not fully charged. $u^{\text{swap},x}_{\text{BSS}}(i,j,t)$ is also a 0/1 variable, which represents whether the battery i of BSS is swapped with a battery of EV j. $Soc^{\text{swap},x}_{\text{BSS,ev}}(j)$ is the SOC of EV j, and $Soc^{\text{swap},x}_{\text{BSS}}(i,j,t)$ is the difference between the SOC of battery i of BSS and the battery of EV j, which is 0 if they are not swapped. Therefore, (2)-(6) are battery swapping constraints of BSS and (7)-(11) are battery charging constraints of BSS, where $P^{\text{ch},x}_{\text{BSS,b}}(i,t)$ is the charge power of battery i in BSS, η^{ch} is the battery charging efficiency, Δt is the time step, $E^{\text{cap}}_{\text{BSS,b}}$ is the capacity of BSS batteries, $P^{\max,\text{ch}}_{\text{BSS,b}}$ is maximal charge power of the battery in BSS and $P^{\max,\text{ch,sum}}_{\text{BSS,b}}$ is maximal total charge power of all batteries in BSS.

$$Soc_{\min} \le Soc^{x}_{\text{EV}}(k,t) \le Soc_{\max}, k \in N_{\text{EV}}(x), t \in S_{t_c}(x) \tag{12}$$

$$\begin{gathered} Soc^{x}_{\text{EV}}(k,t+1) = Soc^{x}_{\text{EV}}(k,t) + P^{\text{ch},x}_{\text{EV}}(k,t)\eta^{\text{ch}}\Delta t / E^{\text{cap}}_{\text{EV}} \\ k \in N_{\text{EV}}(x), t \in S_{t_c}(x) \end{gathered} \tag{13}$$

$$0 \le P^{\text{ch},x}_{\text{EV}}(k,t) \le u^{\text{ch},x}_{\text{EV}}(k,t) P^{\max,\text{ch}}_{\text{EV}}, k \in N_{\text{EV}}(x), t \in S_{t_c}(x) \tag{14}$$

$$P^{x}_{\text{buy}}(t) = \sum_{i \in N_{\text{BSS,b}}(x)} P^{\text{ch},x}_{\text{BSS,b}}(i,t) + \sum_{k \in N_{\text{EV}}(x)} P^{\text{ch},x}_{\text{EV}}(k,t) + P^{x}_{\text{load}}(t), t \in S_{t_c}(x) \tag{15}$$

Equations (12)-(14) refer to EV charging constraints, and (15) is the power balance constraint. $N_{\text{EV}}(x)$ is the set of all EVs, and $Soc^{x}_{\text{EV}}(k,t)$ is the SOC of EV k, $E^{\text{cap}}_{\text{EV}}$ is the capacity of EV batteries, $P^{\text{ch},x}_{\text{EV}}(k,t)$ is the charge power of EV k, $u^{\text{ch},x}_{\text{EV}}(k,t)$ is a 0/1 variable representing whether EV k is charging, $P^{\max,\text{ch}}_{\text{EV}}$ is the maximal charge power of EV, and $P^{x}_{\text{load}}(t)$ is the load power of the residential area.

Constraints of the first stage include (2)-(15). During the proactive defense state, the V2G function of EVs may cause them to charge and discharge repeatedly to achieve the effect of obtaining subsidies and fully charging batteries. However, the proactive defense state lasts for a short time, the subsidies EV can obtain are not much, repeated charging and discharging will affect the battery life, and users' first demand is still to maximize energy storage rather than minimize electricity costs. Therefore, the V2G function of EV is not considered in the proactive defense model.

3.2.2 Second-stage optimization model.

$$\min \sum_{t \in S_{t_c}(x)} p^{x}_{\text{TOU}}(t) P^{x}_{\text{buy}}(t)\Delta t \tag{16}$$

The objective function of the second stage is (16), which is to minimize the cost of electricity consumption, where $p^{x}_{\text{TOU}}(t)$ is time of use (TOU) electricity price.

$$\sum_{t \in S_{t_c}(x)} P^{x}_{\text{buy}}(t)\Delta t \ge G^{1,x}_{2} \tag{17}$$

Constraints of the second stage are the same as those of the first stage, that is, (2)-(15) and (17). $G^{1,x}_{2}$ is the objective function of the first stage, which means the optimization result of the first stage needs to be guaranteed first.

3.2.3 Third-stage optimization model.

$$\min \left(\frac{1}{\left|S_{t_c}(x)\right|} \sum_{t \in S_{t_c}(x)} \left(P^{x}_{\text{buy}}(t) - P^{\text{avg},x}_{\text{buy}} \right)^2 \right)^{1/2} \tag{18}$$

$$P^{\text{avg},x}_{\text{buy}} = \frac{1}{\left|S_{t_c}(x)\right|} \sum_{t \in S_{t_c}(x)} P^{x}_{\text{buy}}(t) \tag{19}$$

The objective function of the third stage is shown in (18)-(19), where $P_{\text{buy}}^{\text{avg},x}$ is the average of total purchased electrical power during the schedule horizon. Since there exists uncertainty in the time of outage, the duration of the proactive defense state may also be at different periods of a day. At the same time, due to the need to store energy as much as possible while meeting the normal electricity demand, the grid load during the proactive defense state is generally higher than the grid load during the economic dispatch state. Therefore, if the time of proactive defense coincides with the peak time of electricity consumption, it is more necessary to consider the third stage to avoid the situation of peak on peak.

$$\sum_{t\in S_{t_c}(x)} p_{\text{TOU}}^{x}(t)P_{\text{buy}}^{x}(t)\Delta t \le G_2^{\text{II},x} \tag{20}$$

In addition to all the constraints of the second stage, it is necessary to consider (20), where $G_2^{\text{II},x}$ is the objective function of the second stage.

3.3 Load restoration model

$$\max \sum_{t\in S_{t_c}(x)} (\lambda_{\text{load}}^{1} P_{\text{load}}^{1,x}(t) + \lambda_{\text{load}}^{2} P_{\text{load}}^{2,x}(t) + \lambda_{\text{load}}^{3} P_{\text{load}}^{3,x}(t)) \tag{21}$$

$$P_{\text{load}}^{x}(t) = P_{\text{load}}^{1,x}(t) + P_{\text{load}}^{2,x}(t) + P_{\text{load}}^{3,x}(t), t \in S_{t_c}(x) \tag{22}$$

The optimization model in load restoration state has only one stage, and the only goal is to restore the power load as much as possible for residential users in the power outage. Therefore, the objective function of this state is (21), where $P_{\text{load}}^{1,x}(t)$, $P_{\text{load}}^{2,x}(t)$ and $P_{\text{load}}^{3,x}(t)$ are respectively the power of the first-level, second-level, and third-level load recovered at time t, and $\lambda_{\text{load}}^{1}$, $\lambda_{\text{load}}^{2}$ and $\lambda_{\text{load}}^{3}$ are the influence coefficients of each level of load. Obviously, the value should be set to $\lambda_{\text{load}}^{1} \gg \lambda_{\text{load}}^{2} \gg \lambda_{\text{load}}^{3}$, so that the model preferentially restores the first-level load.

$$\begin{aligned} &Soc_{\text{BSS,b}}^{x}(i,t+1) = Soc_{\text{BSS,b}}^{x}(i,t) \\ &\qquad + \left(P_{\text{BSS,b}}^{\text{ch},x}(i,t)\eta^{\text{ch}} - P_{\text{BSS,b}}^{\text{dis},x}(i,t)/\eta^{\text{dis}}\right)\Delta t / E_{\text{BSS,b}}^{\text{cap}} \\ &i \in N_{\text{BSS,b}}(x), t \in S_{t_c}(x) \end{aligned} \tag{23}$$

$$0 \le P_{\text{BSS,b}}^{\text{dis},x}(i,t) \le u_{\text{BSS,b}}^{\text{dis},x}(i,t) P_{\text{BSS,b}}^{\text{max,dis}}, i \in N_{\text{BSS,b}}(x), t \in S_{t_c}(x) \tag{24}$$

$$\sum_{i\in N_{\text{BSS,b}}(x)} P_{\text{BSS,b}}^{\text{dis},x}(i,t) \le P_{\text{BSS,b}}^{\text{max,dis,sum}}, t \in S_{t_c}(x) \tag{25}$$

$$u_{\text{BSS,b}}^{\text{ch},x}(i,t) + u_{\text{BSS,b}}^{\text{dis},x}(i,t) \le 1, i \in N_{\text{BSS,b}}(x), t \in S_{t_c}(x) \tag{26}$$

$$\begin{aligned} &Soc_{\text{EV}}^{x}(k,t+1) = Soc_{\text{EV}}^{x}(k,t) \\ &\qquad + \left(P_{\text{EV}}^{\text{ch},x}(k,t)\eta^{\text{ch}} - P_{\text{EV}}^{\text{V2G},x}(k,t)/\eta^{\text{dis}}\right)\Delta t / E_{\text{EV}}^{\text{cap}} \\ &k \in N_{\text{EV}}(x), t \in S_{t_c}(x) \end{aligned} \tag{27}$$

$$0 \le P_{\text{EV}}^{\text{V2G},x}(k,t) \le u_{\text{EV}}^{\text{V2G},x}(k,t) P_{\text{EV}}^{\text{max,dis}}, k \in N_{\text{EV}}(x), t \in S_{t_c}(x) \tag{28}$$

$$u_{\text{EV}}^{\text{ch},x}(k,t) + u_{\text{EV}}^{\text{V2G},x}(k,t) \le 1, k \in N_{\text{EV}}(x), t \in S_{t_c}(x) \tag{29}$$

$$\begin{aligned} &\sum_{k\in N_{\text{EV}}(x)} P_{\text{EV}}^{\text{V2G},x}(k,t) + \sum_{i\in N_{\text{BSS,b}}(x)} P_{\text{BSS,b}}^{\text{dis},x}(i,t) = \\ &\sum_{i\in N_{\text{BSS,b}}(x)} P_{\text{BSS,b}}^{\text{ch},x}(i,t) + \sum_{k\in N_{\text{EV}}(x)} P_{\text{EV}}^{\text{ch},x}(k,t) + P_{\text{load}}^{x}(t), t \in S_{t_c}(x) \end{aligned} \tag{30}$$

$$\begin{aligned} &P_{\text{load}}^{l,x}(t) / P_{\text{load,forecast}}^{l,x}(t) \le P_{\text{load}}^{l,x}(t+1) / P_{\text{load,forecast}}^{l,x}(t+1) \\ &\qquad t \in S_{t_c}(x), l = 1,2,3 \end{aligned} \tag{31}$$

In the load restoration state, the load in the residential area is restored first, so the battery swapping demand of BSS is not considered, and all the batteries of BSS are used to restore the load in the residential area. Therefore, the model does not include constraints (2)-(6). Meanwhile, in order to consider the discharge function of the battery in BSS and the V2G function of EV, some constraints in (7)-(11) and (12)-(15) are replaced by (23)-(30). $P_{\mathrm{BSS,b}}^{\mathrm{dis},x}(i,t)$ is the discharge power of battery i in BSS, η^{dis} is the battery discharging efficiency, $P_{\mathrm{BSS,b}}^{\mathrm{max,dis}}$ is maximal discharge power of the battery in BSS, $P_{\mathrm{BSS,b}}^{\mathrm{max,dis,sum}}$ is maximal total discharge power of all batteries in BSS, $P_{\mathrm{EV}}^{\mathrm{V2G},x}(k,t)$ is the discharge power of EV k through V2G, $P_{\mathrm{EV}}^{\mathrm{max,dis}}$ is the maximal discharge power of EV, $u_{\mathrm{BSS,b}}^{\mathrm{dis},x}(i,t)$ and $u_{\mathrm{EV}}^{\mathrm{V2G},x}(k,t)$ are 0/1 variables respectively representing whether battery i of BSS and EV k discharge.

Considering that the normal operation of residential load requires a continuous power supply, the constraint (31) should be added, so that the restored load can continue to work until the end of the power outage, where $P_{\mathrm{load,forecast}}^{l,x}(t)$ is different level loads are predicted at time t, and l represents the level number.

For the first-level and second-level loads, the model will give priority to their possible restoration, and they are generally much smaller than the third-level load. After considering constraint (31), the total load restored during the optimization horizon may be reduced. However, the reduced restoration load must be the third-level load, and it avoids the situation where the restored third-level load stops restoring in the next period. Therefore, all constraints of the load restoration model are (7), (9)-(10), (12), (14), (23)-(31).

4. Economic and resilient operation strategy process of residential area

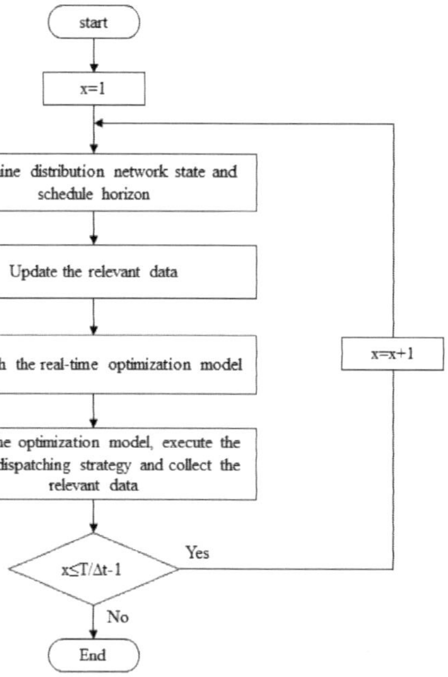

Figure 1. Flow chart of economic and resilient operation strategy for residential area.

In order to take full advantage of the dispatching potential of EVs in residential areas and nearby BSS, and consider the load demand of the residential area and the battery swapping demand of BSS, an economic and resilient operation strategy process is established, as shown in Figure 1, which includes the following steps:

1) Determine the current and future distribution network state, determine the schedule horizon according to the grid state, and select different real-time optimization strategies.

2) Update the relevant data of residential areas and nearby BSS, and predict the data such as residential power load and battery swapping demand of BSS within the schedule horizon.

3) According to the state of the distribution network, establish the real-time optimization model of power dispatching.

4) Solve the real-time optimization model, execute the power dispatching strategy according to the solution results, and collect the relevant data of the residential area and nearby BSS at the end of the schedule horizon.

5) The above steps are repeated at the beginning of each subsequent schedule horizon until the end of the entire optimization horizon to complete a real-time optimization process of power dispatching.

5. Case study

In the case study, the load curve of residential area, the TOU electricity price and EVs' arrival and departure time are referred to [13-14], and the remaining parameters are set as follows: $E^{\text{cap}}_{\text{BSS,b}}$ and $E^{\text{cap}}_{\text{EV}}$ are 75 kWh, $P^{\text{max,ch}}_{\text{EV}}$ and $P^{\text{max,dis}}_{\text{EV}}$ are 7kW, $P^{\text{max,ch}}_{\text{BSS,b}}$ and $P^{\text{max,dis,sum}}_{\text{BSS,b}}$ are 120kW, $P^{\text{max,ch,sum}}_{\text{BSS,b}}$ and $P^{\text{max,dis,sum}}_{\text{BSS,b}}$ are 630kW, Soc_{max} and Soc_{min} are 0.9 and 0.1 respectively, η^{ch} and η^{dis} are 0.9 and 1.1 respectively. The swapping demand curve refers to a certain NIO BSS near the tested residential area from the NIO APP. The proactive defense state lasts for 12 hours, from 12:00 to 24:00.

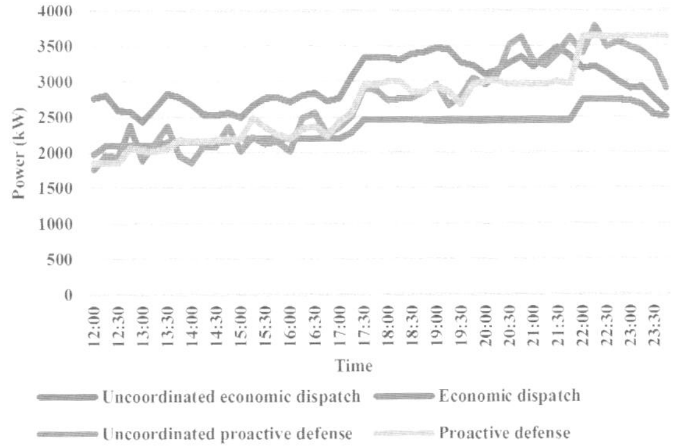

Figure 2. Load curve of economic dispatch and proactive defense state.

Table 1. Evaluation index in proactive defense state.

Total Electricity Cost (¥)		**Peak-valley Difference**	
Uncoordinated model	***Proactive defense model***	***Uncoordinated model***	***Proactive defense model***
17782.51	17649.55	21.34%	20.84%

The proactive defense state lasts for 12h, and after the state, both the residential EV and batteries of BSS are dispatchable resources for load restoration. The comparison of power grid load curve between economic dispatch and proactive defense state is shown in Figure 2. It can be found that the total load curve of the economic dispatch and proactive defense state is quite different that the total load of proactive defense state in the latter half of the time is more than the economic dispatch state. The uncoordinated model of the active defense state also has this tendency, which is to charge the battery as much as possible to cope with the impending power outage. In comparison, the economic dispatching model does not consider the power outage, which will reduce the load restoration in the case of a power outage. It can also be found from Table 1 that the total electricity cost of the optimization model in the proactive defense state decreased by 0.75% compared with that of the uncoordinated model, while the peak-valley difference index increased by 2.32%, showing a small difference.

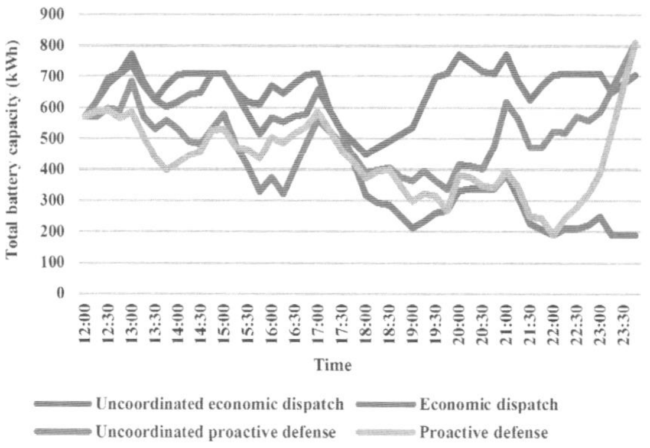

Figure 3. Total battery capacity in economic dispatch and proactive defense state.

Table 2. SOC of each battery in BSS at the last time step of the schedule horizon.

Battery number	Uncoordinated Economic Dispatch	Economic Dispatch	Uncoordinated Proactive Defense	Proactive Defense
1	0.9	0.1	0.9	0.9
2	0.9	0.1	0.9	0.9
3	0.46	0.1	0.9	0.9
4	0.82	0.18	0.9	0.9
5	0.9	0.54	0.9	0.9
6	0.1	0.1	0.9	0.9
7	0.82	0.1	0.9	0.9
8	0.9	0.1	0.9	0.9
9	0.9	0.1	0.9	0.9
10	0.9	0.1	0.9	0.9
11	0.9	0.9	0.9	0.9
12	0.9	0.1	0.9	0.9

As can be seen from Figure 3, the four curves are different, and the two curves in the proactive defense state have the characteristic of charging to reach the upper limit of battery SOC in the last period. The curve of the uncoordinated economic dispatch model has been kept in a high range, while the curve of the optimization economic dispatch model is relatively low, because the latter half of the schedule horizon is in the period of high electricity price, according to the economic dispatch, the load in this part of the time should be transferred to the early morning of low electricity price as far as possible. Obviously, the goals of economic dispatch and proactive defense conflict with each other, which reflects the importance of dividing the state of the power grid. Table 2 shows the SOC of each battery in BSS in the last time step of the schedule horizon, which will be used to restore as much residential load as possible in the next time step, which further illustrates the situation in Figure 3.

As can be seen from Figure 4, in the two models of proactive defense state and the uncoordinated model of economic dispatch, the residential EVs with V2G function will let batteries charge as much as possible during last few time steps of the schedule horizon, while the optimization model of economic dispatch state is on the contrary, and even make profits from V2G discharge in the evening peak period. As a result, there is a difference of more than 50% in the total amount of power that can be used in the following load restoration state, which will seriously affect the dispatchable ability of EVs during power outages.

The load restoration results after the proactive defense state are compared with the load restoration results after the economic dispatch state. The load restoration state lasts for 2h, that is, the power outage period is 12:00-14:00.

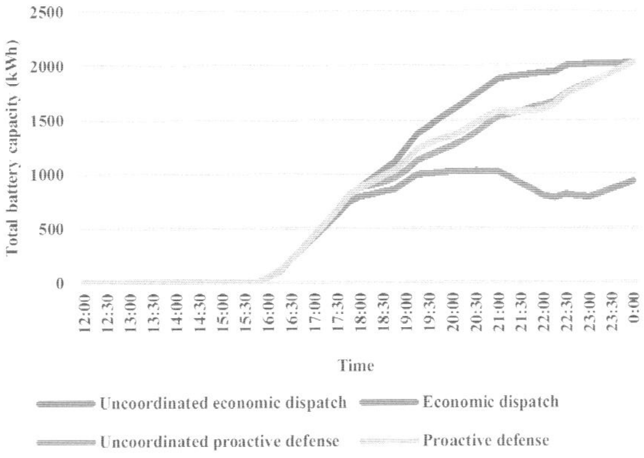

Figure 4. Total battery capacity curve of residential EVs with V2G function.

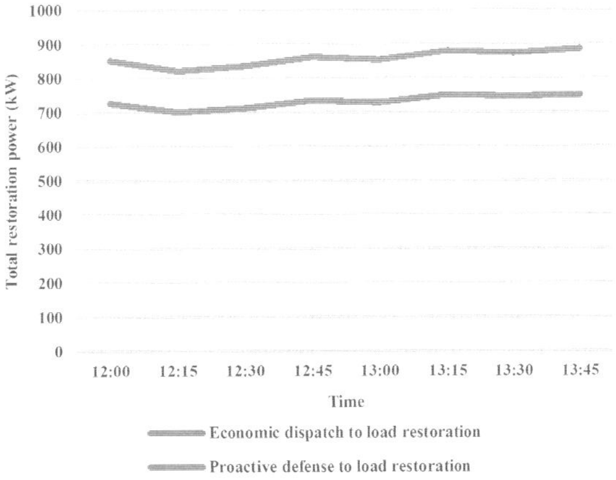

Figure 5. Restoration power curves after different states.

In Figure 5, the load recovered after the economic dispatch state is lower than the load recovered after the proactive defense state, which can also be seen from the above content. In the load restoration stage, the recovered load will continue to work in a relatively stable state without large fluctuations due to constraint (31).

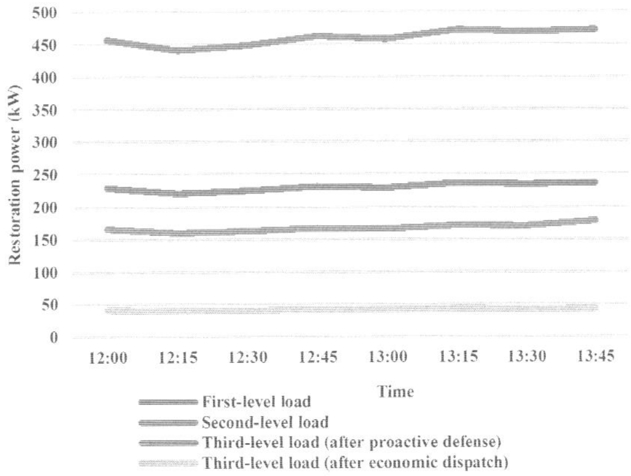

Figure 6. Restoration power curves of different levels after different states.

In the two models of load restoration state, both the first-level and the second-level load can be fully restored, as shown in Figure 6, which indicates that the available power in the tested system is sufficient to support the uninterrupted use of important loads.

6. Conclusion

The paper proposes an economic and resilient operation strategy of a residential area based on nearby BSS to reduce the total electricity cost of users and reduce the peak-valley load difference of the power grid. Because the demand priority of the distribution network is different in different states, the paper first defines three distribution network states and gives a schedule horizon according to different states. Then, the multi-stage optimization models under different states are established, and the specific objective functions and corresponding constraints are given. Finally, in the case study, the validity of the proposed models is studied through the test system. The case study proves the superiority of the proactive defense state over the economic dispatch state for the subsequent load restoration state, and also compares the difference between uncoordinated models and optimization models.

Acknowledgment

The work described in this paper was supported by the Science and Technology Project of State Grid Corporation of China, “Research on Key Technologies for Mining the Value of Adjustable Load Resources and Designing Business Models(1300-202157404A-0-0-00)”.

References

[1] X. W. Yan, Q. A. Wang and J. D. Lu, “Microgrid energy scheduling with electric vehicles and flexible loads,” Power System Protection and Control, vol. 51, no. 17, pp.69-79, September 2023.
[2] Y. Peng and X. M. Bai, “What EV users say about policy efficacy: Evidence from Shanghai,” Transport Policy, vol. 123, pp.16-26, 2023.
[3] K. Sun, W. G. Wen, and J. J. Chen, “A resilience enhancement strategy for multi-microgrid in extreme scenarios based on electric vehicle,” Power System Protection and Control, vol. 51, no. 24, pp.53-65, December 2023.
[4] L. Calearo, A. Thingvad and K. Suzuki, “Grid loading due to EV charging profiles based on pseudo-real driving pattern and users’ behavior,” IEEE Transactions on Transportation Electrification, vol. 5, no. 3, pp.683-694, September 2019.
[5] F. Wu, J. Yang, and X. Zhan, “The online charging and discharging scheduling potential of electric vehicles considering the uncertain responses of users,” IEEE Transactions on Power Systems, vol. 36, no. 3, pp.1794-1806, May 2021.
[6] S. R. Revankar and V. N. Kalkhambkar, “Grid integration of battery swapping station: A review,” Journal of Energy Storage, vol. 41, pp.102937, September 2021.
[7] Y. B. Chen, W. R. Wang and Y. F. Yang, “Grid dispatching optimization based on electric vehicle charging load demand forecast,” Electrical Engineering, vol. 5, pp.31-33, May 2023.
[8] H. Hou, J. Y. Tang and Y. F. Wang, “Long-time scale charging and discharging scheduling of electric vehicles under joint price and incentive demand respons,” Automation of Electric Power Systems, vol. 46, no. 15, pp.46-55, August 2022.
[9] S. N. Wang and S. B. Yang, “A coordinated charging control strategy for electric vehicles charging load in residential area,” Automation of Electric Power Systems, vol. 40, no. 4, pp.71-77, February 2016.
[10] B. Sun, X. Tan, and D. H. K. Tsang, “Optimal charging operation of battery swapping and charging stations with QoS guarantee,” IEEE Transactions on Smart Grid, vol. 9, no. 5, pp. 4689-4701, September 2018.
[11] H. Wu, G. K. H. Pang and K. L. Choy, “An optimization model for electric vehicle battery charging at a battery swapping station,” IEEE Transactions on Vehicular Technology, vol. 67, no. 2, pp. 881-895, February 2018.
[12] H. Ko, S. Pack and V. C. M. Leung, “An optimal battery charging algorithm in electric vehicle-assisted battery swapping environments,” IEEE Transactions on Intelligent Transportation Systems, vol. 23, no. 5, pp. 3985-3994, May 2022.
[13] Y. P. Li, H. Su, and Y. Zhou, “Two-stage real-time optimal electricity dispatch strategy for urban residential quarter with electric vehicles’ charging load,” Energy, vol. 268, pp. 126702, April 2023.
[14] X. Yang, W. Lu and W. C. Yu, “Optimal dispatching and control strategies for residential load of intelligent communities,” Power System Protection and Control, vol. 51, no. 21, pp.22-34, November 2023.

Journal of Physics: Conference Series **3031** (2025) 012006 doi:10.1088/1742-6596/3031/1/012006

Simulation and analysis of lithium-ion batteries thermal runaway during overcharge

Jiuchun Jiang[1], Jinqiao Du[2], Wenwei Zhan[1], Jie Tian[2], Lipeng Yang[1], Bo Rao[2], and Xiaoguang Yang [1*]

[1] Shenzhen Automotive Research Institute, Beijing Institute of Technology, Shenzhen 518118, Guangdong, China
[2] Shenzhen Power Supply Co., Ltd., Guangdong Provincial Key Laboratory of Source-Grid-Load-Storage Interactive Collaborative Technology (No. 2024B1212020004), Shenzhen 518000, Guangdong, China

[*] Corresponding author's e-mail: wxt196322@126.com

Abstract. To solve the problem of temperature distribution and voltage variation of lithium-ion batteries under overcharge conditions, an electro-thermal coupling model considering side reactions was established, and the overcharge simulation of square soft packet lithium-ion batteries under different current rates was carried out. The results show that the changes in battery voltage and temperature under the condition of overcharge have three stages, and the difference between different stages is significant. The thermal effect caused by high-rate current is more obvious, the battery temperature distribution is more uneven, and the area of the high-temperature region is smaller. The heat transfer path is the same and the time is different under different magnifications. Compared with the general model, the modeling strategy presented in this paper has a better adaptability to the temperature rise stage before the battery's internal short circuit occurs.

1. Introduction

Energy storage is a crucial component of the energy internet and a key technology supporting the construction of a new power system. Electrochemical energy storage, particularly lithium-ion battery-based systems, is highly favored due to its rapid response, excellent scalability, and flexible regulation capabilities. With the introduction of the "Dual Carbon" goals, the National Development and Reform Commission and the National Energy Administration have successively issued a series of policies, leading to rapid development in the lithium-ion battery energy storage industry. It is projected that by 2025, the installed capacity of new energy storage systems will reach 30 million kilowatts [1]. The advantages of lithium-ion batteries, such as high energy density, low self-discharge, memorylessness, and long life, have greatly promoted the development of electrochemical energy storage. However, lithium-ion battery safety accidents characterized by thermal runaway have occurred occasionally [2-5].

Overcharge is one of the most common reasons for battery thermal runaway, which is difficult to avoid to some extent. When the battery is overcharged for a long time due to internal battery faults or improper charging behavior and no corresponding measures are taken, the battery system will be seriously damaged, leading to the occurrence of fire, explosion, and other dangerous accidents [6].

For the thermal runaway phenomenon caused by overcharge, a lot of experiments and simulations have been carried out. Based on the law of SOC changing with the voltage, Chuang Qi established the heat source equation of lithium-ion batteries and proposed the overcharging-thermal coupling model by combining numerical simulation and experiment, which predicted the critical point of thermal runaway. However, this model only considered the influence of the overcharged current rate on the temperature field distribution of the battery, and could not calculate the thermal effect of the overcharged battery based on the charging time [7]. Through the analysis of the experimental phenomenon, Yanxian Sun has calculated the heat generated during the overcharge of the cell by using the internal resistance method and the internal short circuit model. Subsequently, they built a battery module overcharge simulation model based on the heat diffusion equation between the cells and gave the starting time of thermal runaway within a certain range. Although this model can simulate the temperature distribution of the overcharged battery within a certain range, no further research has been conducted on the thermal runaway of the overcharged battery [6]. Jiana Ye has defined the semi-failure state of a lithium battery and established an overcharged internal resistance model under the adiabatic condition to study the changes in the internal resistance of the battery during overcharge, to predict the occurrence time of the semi-failure state of the battery. These studies have provided a basis for the early warning of battery thermal runaway, but the conclusions are not universal because only one type of battery was selected in the relevant experiments and the objects were relatively single [8]. Based on the above research and supported by the actual physical laws of the battery, it is of great significance to select an appropriate overcharge standard for the simulation of the lithium-ion power battery, which can not only reduce the experimental cost but also provide a relevant basis for the actual experiment.

In this paper, the author established a square ternary lithium-ion battery model and coupled the side reactions of lithium batteries under high temperatures with the basic electrochemical model by using COMSOL Multiphysics. The temperature distribution and composition of the lithium-ion battery under different overcharge scales were obtained, and the thermal runaway phenomenon of the battery under the overcharge condition was analyzed in combination with the variation law of the voltage.

2. Establishment of model

2.1 Electrochemical-thermal coupling model of lithium batteries

In the coupling model, the electrochemical module adopts the P2D model proposed by Newman et al. [9]-[13], which is based on the conservation of quality, charge, and energy. The Butler-Volmer equation and Fick's law are used to characterize the electrochemical phenomenon of the electrode-electrolyte interface and the transport process of solid lithium in the battery, which can reflect the charge and discharge behavior of the battery under normal conditions to a large extent. At the same time, when using this model, the following basic assumptions are defaulted: (1) The uniformly distributed particles inside the solid electrode are spherical, and their transmission process satisfies Fick's second law; (2) The charge and mass transfer rules in the electrolyte follow the concentrated solution theory; (3) The material inside the battery is isotropic and its specific heat capacity and thermal conductivity is constant. The polarization heat, reversible heat, and ohmic heat in electrochemistry are used as the heat source of the lithium-ion battery under normal working conditions in the heat module. Combined with the set initial value and boundary conditions, the calculated temperature can be transferred back to the electrochemistry module to realize the coupling and calculation of the temperature in the battery system.

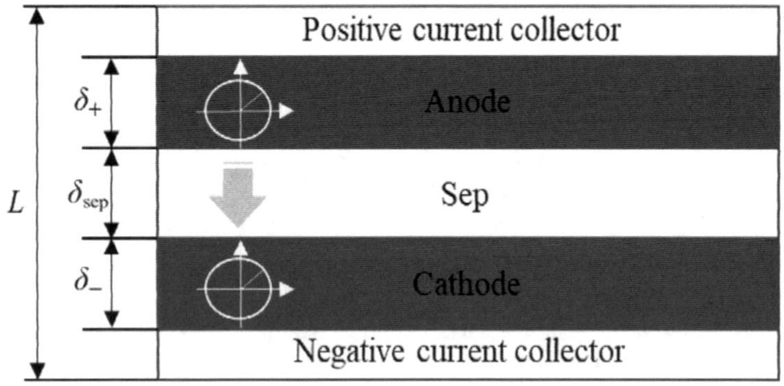

Figure 1. Lithium-ion battery model.

Journal of Physics: Conference Series **3031** (2025) 012006 doi:10.1088/1742-6596/3031/1/012006

Since the P2D quasi-two-dimensional model proposed by Newman et al. (see Figure 1) is based on the theory of porous electrode and concentrated solution, the control equations in the model include electrode dynamics, solid phase diffusion of porous electrodes, and electrolyte phase diffusion of porous electrodes. In addition, the model also introduces the heat generation process of side reactions according to energy conservation.

The ternary lithium-ion power battery adopts porous solid-phase electrodes. The contact between these electrodes and the electrolyte provides a path for the ion transmission from one pole of the battery to the other pole. However, the electrons generated by the electrode reaction also need to cross the electrode-electrolyte interface to achieve the charge balance of the battery. The overvoltage and current exchange density together constitute the Butler-Volmer dynamics expression, and the four field variables $c_{s,e}(x,t)$, $c_e(x,t)$, $\phi_s(x,t)$, and $\phi_e(x,t)$ are coupled. Thus, the electron conservation and lithium-ion conservation in the phase diffusion of the solid phase and electrolyte of the porous electrode can be obtained, and the corresponding boundary conditions are added to form the pseudo-two-dimensional model of lithium-ion battery. The specific mathematical expressions and boundary conditions are shown in Figure 2.

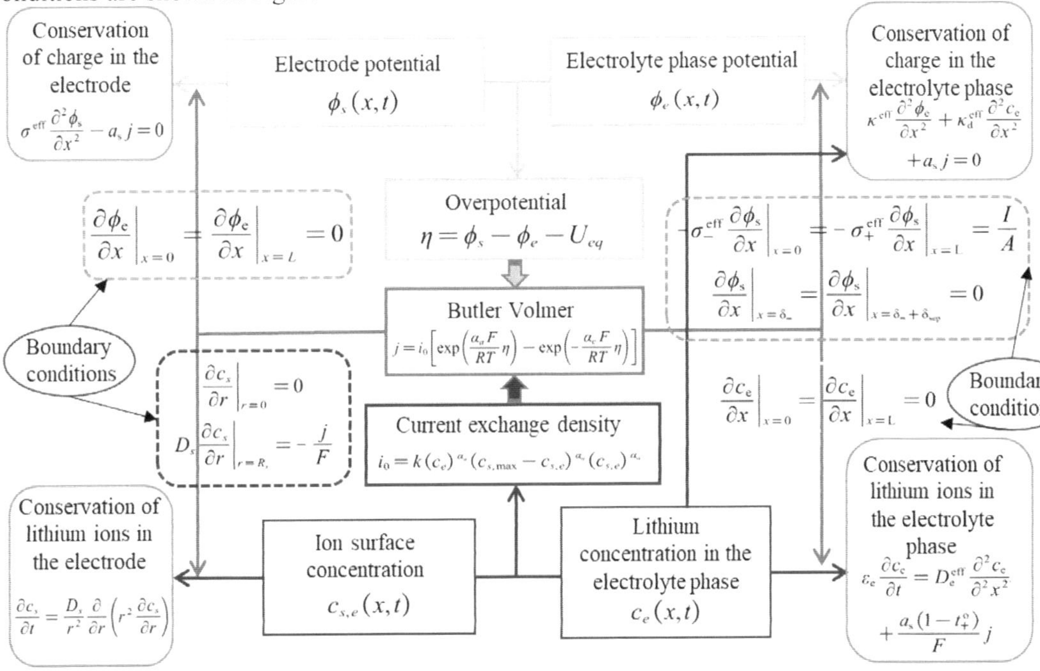

Figure 2. Mathematical model of lithium-ion battery.

2.2 Heat generation process of lithium-ion battery under overcharge

Relevant experiments show that lithium-ion batteries will accumulate heat during charging, and most of this heat is caused by the loss of ohmic heat of the battery, and polarization heat occupies a small part [14]. After a period of overcharging, the heat accumulation will be more and more serious. If there is no good heat dissipation condition at this time, when the temperature inside the battery continues to rise to a certain value, some side reactions will begin to occur. Therefore, in the simulation of battery overcharging, polarization heat Q_p, Joule heat Q_j, and side reaction heat generation Q_s were mainly considered [15].

From the beginning of the battery overcharge to the temperature reaching 100°C, the battery's heat generation power changes with time, as shown in Figure 3.

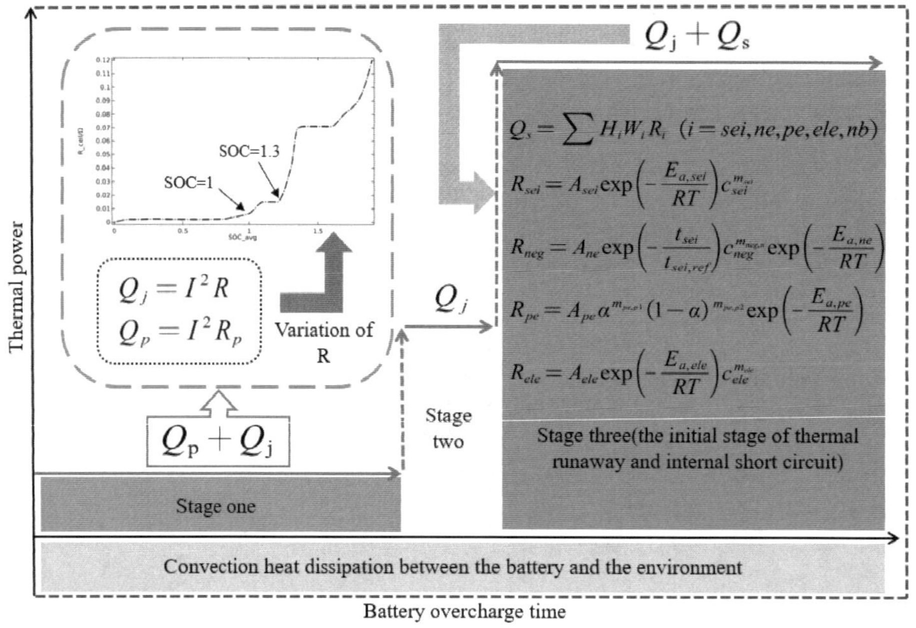

Figure 3. Chart of thermal power in the overcharge stage of battery.

As the temperature gradually accumulates, the battery may change physical structure such as expansion and deformation, so joule heat is the main heat source during the entire temperature rise process. The increasing internal resistance of the battery charging brings a significant thermal effect, which makes the temperature rapidly rise. In the later stage of charging, side reactions also provide part of the heat, and there are four main effects: SEI decomposition reaction, anode-electrolyte reaction, cathode-electrolyte re-action, and electrolyte decomposition reaction [16-19].

SEI decomposition reaction: The negative electrode of the Li-ion power battery has a layer called SEI. The SEI is composed of a stable layer and a meta-stable layer. Its insulating structure mainly plays a protective role in preventing the negative electrode material from reacting with the electrolyte. When the temperature of the battery reaches 90-120°C, SEI will decompose due to instability, thereby releasing heat.

Anode-electrolyte reaction: taking the instability of the SEI into account, when the temperature exceeds 120°C, the SEI cannot protect the negative electrode. The lithium embedded in the negative electrode will react with the electrolyte to release heat.

Cathode-electrolyte reaction: in an oxidized state, the positive electrode material can directly react with the electrolyte or the oxygen generated by the exothermic decomposition of the positive electrode active material reacts with the electrolyte. In any case, the chemical reduction of the positive electrode active material with electrolyte is highly exothermic.

Electrolyte decomposition reaction: under gradually increasing temperature, the electrolyte inside the battery will react with other components and lose a part. It will undergo thermal decomposition at different temperatures. These processes will all produce heat.

In addition, for the heat dissipation process, radiation heat dissipation is ignored due to the small size of the battery. The convection heat dissipation between the battery and the surrounding environment is considered.

2.3 Selection of experimental conditions for overcharging

To standardize the use and testing of lithium batteries, a series of standards have been issued at home and abroad [20]. Some are selected to analyze and determine the experimental conditions for simulation (Table 1).

Journal of Physics: Conference Series **3031** (2025) 012006 doi:10.1088/1742-6596/3031/1/012006

Table 1. Lithium-ion battery overcharging experiment conditions.

Serial number	Test standard name	Overcharge standard
1	OCT 743-2006 [21]	3C charge to 5 V or 250% SOC or 9C charge to 10 V
2	GB/T 31485-2015 [22]	1C charge to 1.5×Vmax or 200% SOC
3	GB/T 31467.3-2015 [23]	1C charge to 1.2×Vmax or 130% SOC
4	IEC 62660-2-2010 [24]	1C charge to 2×Vmax or 200% SOC
5	IEC 62113-2012 [25]	1C or maximum charging current to 250% SOC

The overcharge standards listed in the table cover three factors: the charging current rate, the charging depth, and the cut-off voltage. Taking the simulation capability and accuracy of the software into account, the implementation rules of GB/T 31467.3-2015 were used for this experiment to enable a better comparative analysis of the heat change during battery overcharge. Based on this standard, the temperature rise corresponding to 1C, 2C, and 3C current multipliers was tested.

2.4 Physical modeling of experimental battery

The anode material of the battery is lithium manganese oxide (LMO). The cathode material is graphite. The electrolyte is LiPF6 in 3:7EC: EMC. The positive and negative fluid collectors are aluminum and copper respectively. The battery size is 200 mm×100 mm×15 mm. The geometry of the model was dissected by using a free quadrilateral mesh. The physical model is shown in Figure 4.

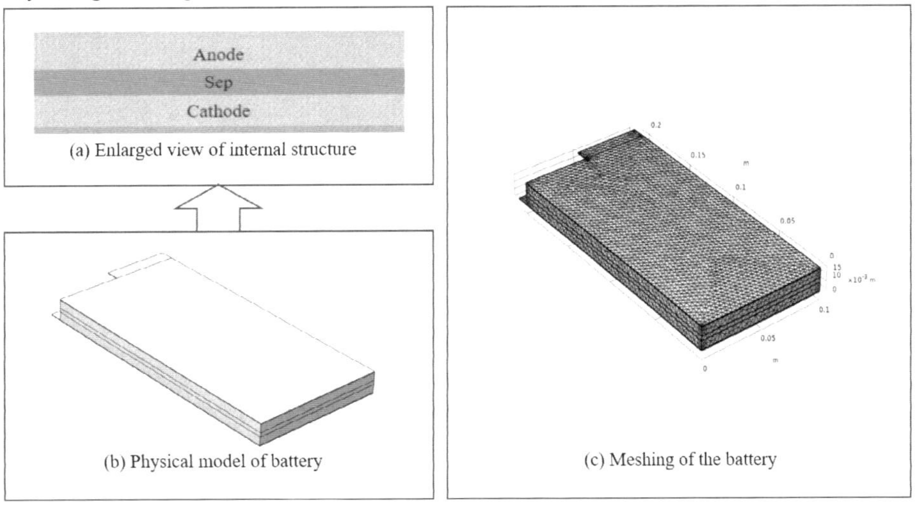

(a) Enlarged view of internal structure
(b) Physical model of battery
(c) Meshing of the battery

Figure 4. The physical model with mesh sectioning.

The related side reaction parameters are shown in Table 2. The specific initial values of relevant parameters such as C_{sei}, C_{neg}, C_{pos}, C_{ele}, α_0, and $T(0)$ can be set to 0.15, 0.75, 1, 0.04, 0.04, and 298 K respectively. The convective heat transfer coefficient between the battery and the external environment is set to 20 W/(m^2•K).

Table 2. High-temperature side reaction parameters of ternary lithium battery.

Parameter	①	②	③		④
ΔH_x(J/kg)	2.57×10^5	1.74×10^6	1.46×10^5	1.46×10^5	1.55×10^5
Δm_x(kg/m^3)	261.63	261.63	63.3	63.3	321.5
A_x(1/s)	1.667×10^{15}	2.5×10^{13}	1.05×10^{11}	6.64×10^{13}	5.14×10^{25}
$E_{a,x}$(J/mol)	1.35×10^5	1.35×10^5	1.14×10^5	1.58×10^5	2.74×10^5

3. Results and discussion

To standardize the use and testing of lithium batteries, a series of standards have been issued at home and abroad [20]. Some are selected to analyze and determine the experimental conditions for simulation (Table 1).

3.1 Changes in voltage and temperature during overcharging

Figure 5 shows the temperature and voltage change curve when the battery is overcharged to 130% SOC at the rate of 1C. From this curve, when the almost fully charged battery continues to be energized, the battery temperature increases normally. From the composition of the heat, it can be known that the battery temperature rise under normal conditions mainly comes from Joule heat and polarization heat. Joule heat depends on the ohmic internal resistance of the battery. The polarization heat is caused by the obstruction of the internal polarization resistance to the current. The ohmic internal resistance is related to the structure, assembly, size, and other factors of the battery itself. The polarization internal resistance is not only affected by the charging speed but also closely related to the overpotential in the battery. When the battery is overcharged, with the reduction of the content of lithium in the positive electrode, the de-lithium-embedded reaction in the battery becomes more and more difficult. The obstruction to the current is further enhanced [26][27]. Therefore, the battery temperature will gradually rise along with the process of overcharging. Meanwhile, oxidizing gases are also generated inside the battery due to the destruction and collapse of certain microstructures [28]. These gases react with the electrolyte to release heat and expand the battery pack, which further intensifies the exothermic reaction inside the battery. Until the diaphragm is melted or pierced by dendrites [29-32], the positive and negative electrodes contact indirectly.

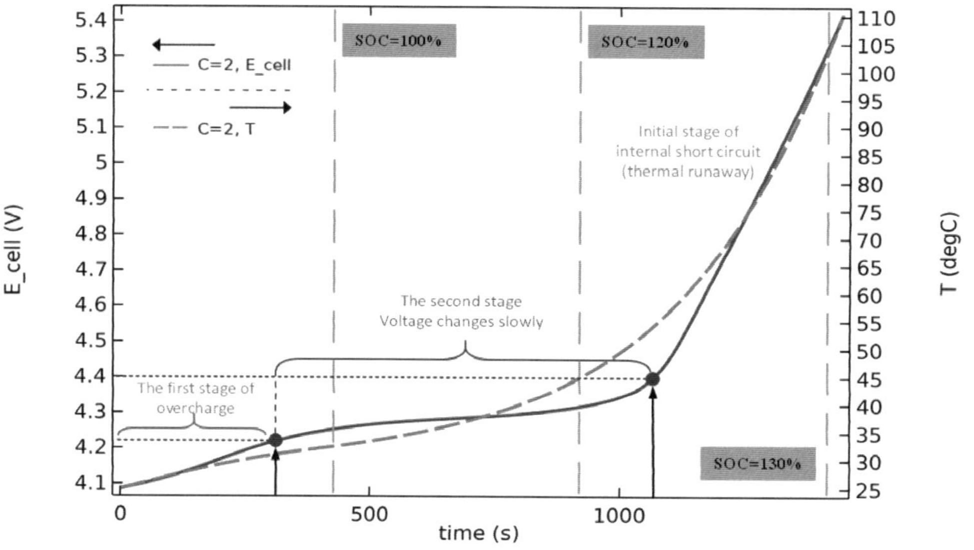

Figure 5. Temperature and voltage change curves of the battery before the occurrence of internal short circuits.

In the initial stage of overcharge, the battery is not fully charged, so the speed of voltage increase is relatively normal. After reaching the charging cutoff voltage of 4.2 V, the lithium intercalation reaction of the negative electrode continues, but the lithium delithium of the positive electrode cannot completely reach the negative electrode. Compared with the first stage, the voltage changes slowly, but the temperature rises more quickly in this stage. In the last stage, the battery voltage has an obvious transient and the temperature rises more rapidly. It reflects the beginning of the lithium precipitation reaction of the graphite negative electrode. Dendrites begin to grow, and the negative potential gradually deviates due to excessive lithium insertion in the negative electrode [33], which results in a sharp change in voltage at this stage.

Journal of Physics: Conference Series **3031** (2025) 012006 doi:10.1088/1742-6596/3031/1/012006

3.2 Variations in voltage and temperature caused by charging current at different rates

The external parameters of the battery were successively calculated under 1C, 2C, and 3C current overcharge, as shown in Figures 6-7.

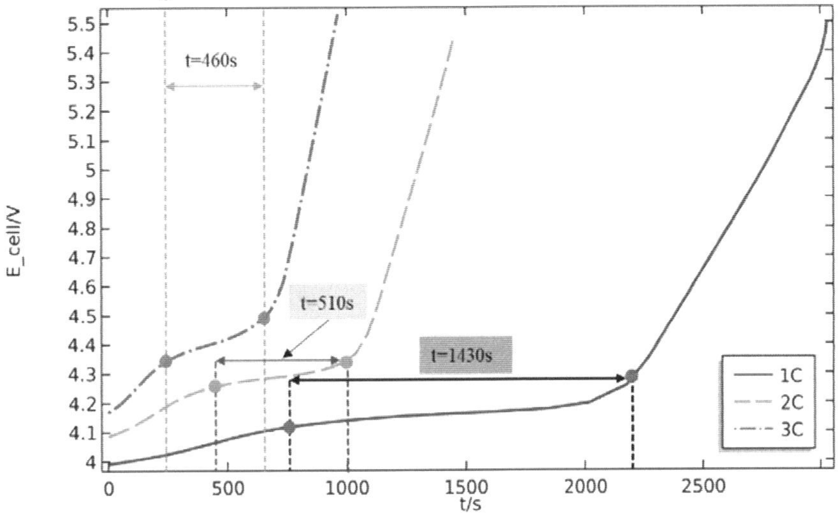

Figure 6. Variation of battery voltage at different ratios.

For different current ratios, voltage and temperature change obviously, indicating that different overcharge currents have an important impact on the whole battery thermal runaway process. This is because the heat produced before the thermal runaway is mainly from Joule heat and polarization heat, both of which are closely related to current. The higher current ratio increases the conversion of internal chemical energy to thermal energy in the battery, makes the polarization phenomenon more serious, and accelerates the process of thermal runaway.

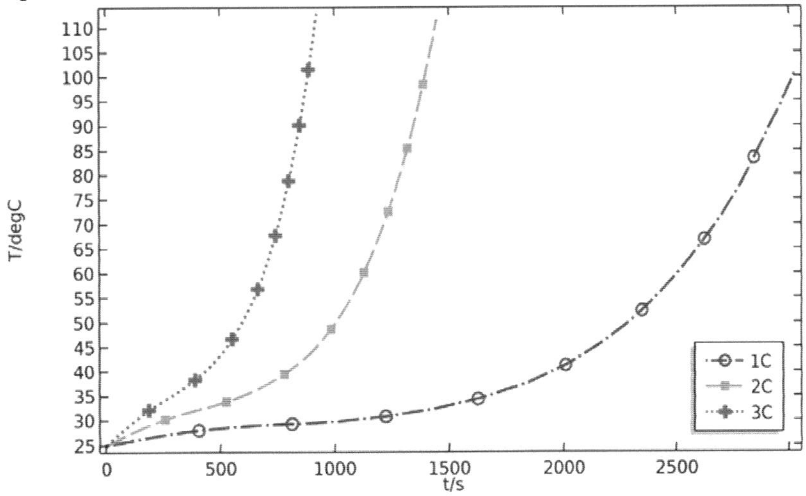

Figure 7. Changes in battery temperature at different ratios.

3.3 Temperature distribution at different current ratios

The battery temperature distribution during the whole charging process can be obtained by simulation. Figure 8 shows the battery thermal effect caused by the 1C-3C current. With the battery charging process, the main heat source gradually transfers from the positive electrode ear to the battery core. And as time goes on, the heat gradually spreads to the edges.

The battery material properties are not uniform, which is not only related to the manufacturing process but also related to the change in the temperature field. So, it is more practical to compare the changing trend of the battery temperature under the same conditions. In the case of 1C current, the total simulation time is 3, 000 s. The temperature distributions at 110%, 120%, and 130% SOC are intercepted. In the case of 2C and 3C currents, the total simulation time is 1, 450 s and 960 s, respectively. Three interception times remain the same as 1C.

Horizontal and longitudinal comparison shows that with the increase of charging current ratios, the battery temperature distribution is more and more uneven. The high-temperature part is all transferred from the pole ear to the right side of the cell center. At the beginning of overcharging, the electrode ear, as a receiving part, has a more obvious change in the current. Moreover, the electrode ear is small in volume and thermal physical property parameters change very small. As a result, its temperature distribution is uniform and heat transfer is faster.

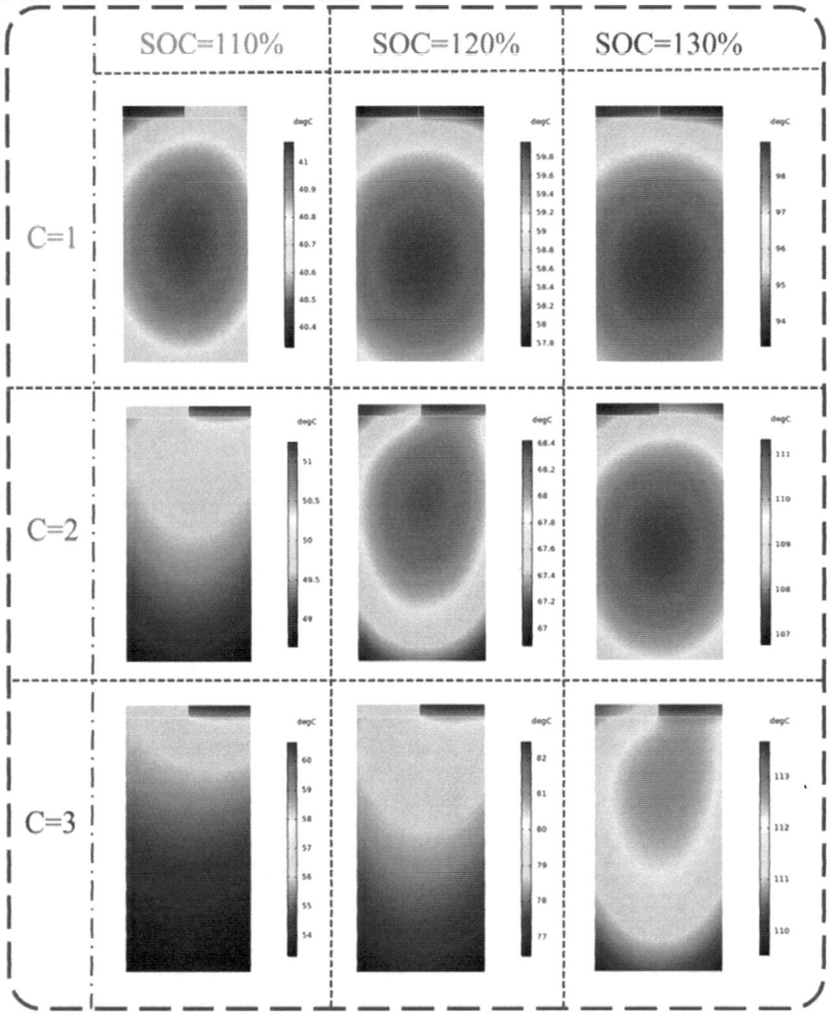

Figure 8. Temperature distribution at different overcharge multipliers.

However, with the deepening of the battery reaction, the heat generation at the electrode ear becomes very small, and the main heat source is transferred to the lower side. Meanwhile, the higher current ratio brings the faster heat generation rate, which makes the battery temperature distribution difference become bigger gradually. In addition, the heat generation of the positive electrode ear is higher than that of the negative one. The high-temperature area will be smaller and smaller, indicating

Journal of Physics: Conference Series **3031** (2025) 012006 doi:10.1088/1742-6596/3031/1/012006

that the thermal effect brought by the high current ratio is more concentrated. When heat dissipation is much less than its production in a short period, the battery temperature rise will be faster. Before entering the internal short circuit stage, the temperature has reached a certain high value. Under such conditions, the battery side reaction will provide a certain energy for the subsequent temperature jump. The occurrence of the four types of side-reaction can be inferred according to the dimensionless content of the corresponding substances inside the battery. The specific results are shown in Figure 9 (3C ratio).

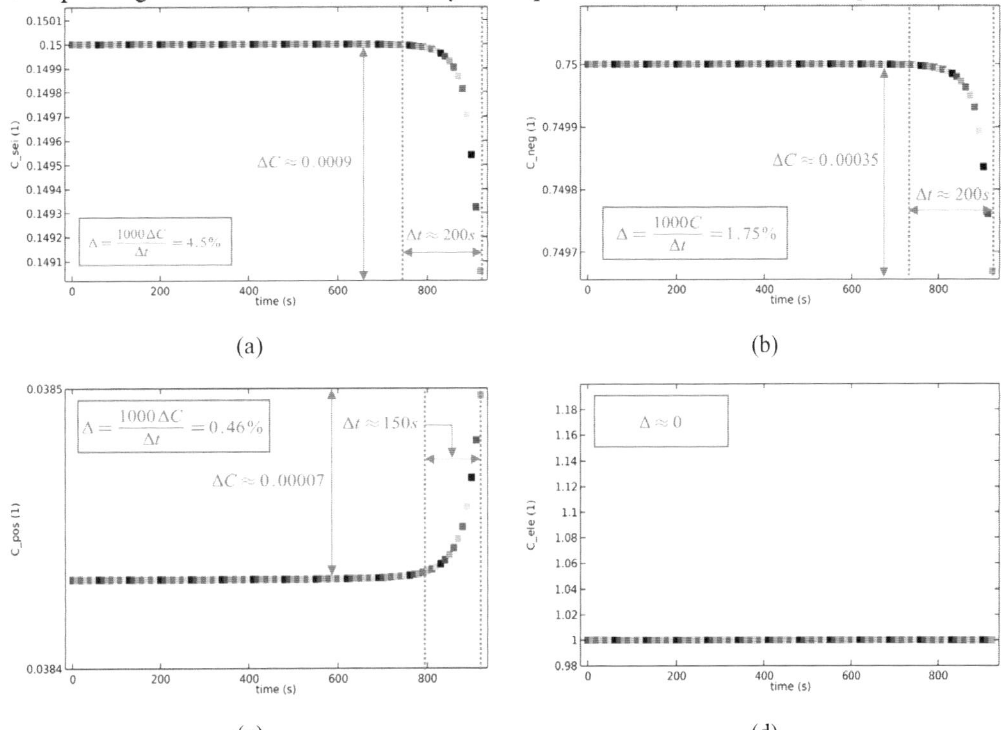

Figure 9. The occurrence of side reactions.

When the reaction is carried out for 800 s, the SEI decomposition and anode-electrolyte reaction almost occur at the same time when the temperature is 90°C. After 100 s, the cathode-electrolyte reaction gradually begins, and the average temperature of the battery has reached about 100°C. In addition, the heat generated by the above three reactions rises continuously.

4. Conclusion

For the lithium-ion battery temperature distribution and voltage variation under overcharge conditions, an electric-thermal coupling model was established. The overcharge simulation under different current ratios was carried out. The simulation results show that:

(1) For the study of lithium-ion batteries under complex and dangerous conditions such as overcharging, the numerical simulation method can effectively simplify the experimental design based on keeping the corresponding physical law unchanged. It is a good reference for difficult numerical sampling.

(2) From the beginning of overcharge to the temperature rising to 100°C, the battery voltage and temperature change can be roughly divided into three stages. The first stage is a short-term overcharge period, in which the voltage and temperature keep increasing normally. When the voltage reaches the critical value, the battery enters the second stage, in which the voltage changes slowly but the

temperature rises quickly. Finally, when the battery voltage begins to rise rapidly, the battery enters the third stage. During this period, side reactions begin to occur and the temperature soon jumps to 100°C.

(3) There are great differences in the thermal effects caused by the overcharge current with different rates, which are reflected in the temperature distribution and voltage variation. The higher charging ratio causes the more obvious temperature rise of the battery and the corresponding temperature distribution is more uneven.

Due to the complex internal physical structure of the battery and the difficulty in quantitative chemical reactions observation, most of the current research starts from the external parameters to predict the state of the battery under different conditions. When the lithium-ion battery is overcharged, it will form lithium dendrites on the surface of the negative electrode because of the difficulty in embedding lithium. When the dendrite penetrates the membrane, the battery will have a serious internal short circuit [34][35]. Therefore, when the overcharge temperature reaches more than 100°C, the accuracy of the simulation model will be reduced, which is one of the defects of the model established in this paper. It will be more applicable to establish the corresponding model for the temperature and dendrite growth of lithium-ion batteries in the delimit stage.

Acknowledgments

This work is financially supported by the China Southern Power Grid Science and Technology Project (090000KC23120055) and Shenzhen Science and Technology Program (KJZD20230923115212026, JCYJ20220530172601003).

References

[1] Wang Zhenpo, et al. "Effects of short-term over-discharge cycling on the performance of commercial 21, 700 lithium-ion cells and the identification of degradation modes". Journal of Energy Storage 35, (2021). Doi: 10.1016/J. EST. 2021. 102257.
[2] Peng Zhang, Jiajia Zhu, Miao Wang, Nobuyuki Imanishi, and Osamu Yamamoto. "Lithium dendrite suppression and cycling efficiency of lithium anode". J. Electrochemistry Communications, 2018.
[3] Michael Naguib, Srikanth Allu, Srdjan Simunovic, Jianlin Li, Hsin Wang, and Nancy J. Dudney. "Limiting internal short-circuit damage by electrode partition for impact-tolerant Li-ion batteries". J. Joule. 2018 (1).
[4] Mingxuan Zhang, Minggao Ouyang, Languang Lu, Xiangming He, Xuning Feng, Lishuo Liu, and Xiaoyi Xie. "Battery internal short circuit detection". J. ECS Transactions. 2017 (11).
[5] Zhang Mingxuan. "Research on the internal short circuit problem of the vehicle power battery system", D. Tsinghua University, 2018.
[6] Sun Yanxian. "Study on thermal simulation and safety improvement method of power lib with ternary material", D. Harbin Institute of Technology, 2020 (in Chinese).
[7] Wang Congjie, Zhu Yanli, Fan Xuekun, Qi Chuang, and Gao Fei. "Mathematical model for thermal behavior of lithium-ion battery pack under overheating". J. Applied Thermal Engineering, 2021, 191.
[8] Jiana Ye, Haodong Chen, Qingsong Wang, Peifeng Huang, Jinhua Sun, and Siuming Lo. "Thermal behavior and failure mechanism of lithium-ion cells during overcharge under adiabatic conditions". J. Applied Energy, 2016, 182.
[9] Marc Doyle, Thomas F. Fuller, and John Newman. "Modeling of galvanostatic charge and discharge of the lithium/polymer/insertion cell". J. Journal of the Electrochemical Society, 2019, 140 (6).
[10] Guoying Chen, Karen E., Thomas-Alyea, John Newman, and Thomas J. Richardson. "Characterization of an electroactive polymer for overcharge protection in secondary lithium batteries". J. Electrochemical Acta, 2005, 50 (24).
[11] John Christensen and John Newman. "Erratum: a mathematical model for the lithium". J. Journal of the Electrochemical Society, 2005, 152 (3).
[12] Wu T. & Schmidt A. (2025). Real-time thermal-electrochemical monitoring of solid-state batteries using embedded sensors. Advanced Energy Materials, 15 (7), 2, 203, 456. https://doi.org/10.1002/aenm.202503456.
[13] Liu B., Zhang F., & Patel S. (2024). Machine learning-driven optimization of fast-charging protocols with coupled thermal constraints. Nano Energy, 98, 107, 892. https://doi.org/10.1016/j.nanoen.2024.107892.
[14] Kumar R. & Lee J. (2025). 3D-printed battery architectures: Coupled thermal and electrochemical performance analysis. ACS Energy Letters, 10 (3), 789-795. https://doi.org/10.1021/acsenergylett.5c00012.
[15] Chen Z., Wang Q., & Kim Y. (2024). Thermal runaway suppression in high-voltage Li-ion batteries via phase-change composites. Energy Storage Materials, 70, 102, 567. https://doi.org/10.1016/j.ensm.2024.102567.
[16] Gómez A. & Srinivasan V. (2025). Degradation-aware electro-thermal modeling for second-life battery systems. Journal of Power Sources, 612, 234, 567. https://doi.org/10.1016/j.jpowsour.2025.234567.

[17] Zhang L. & Ito M. (2024). Multiphysics simulation of lithium-sulfur batteries under low-temperature cycling. Electrochimica Acta, 480, 12, 9876. https://doi.org/10.1016/j.electacta.2024.129876.
[18] Nguyen H. & Gupta P. (2025). AI-enhanced thermal management for electric vehicle battery packs. In Proceedings of the 2025 IEEE International Conference on Sustainable Energy Technologies (pp. 45-52). IEEE. https://doi.org/10.1109/ICSET.2025.00012.
[19] Wang X. & Zhang Y. (2024). Dynamic coupling of electrochemical-thermal models for wireless charging applications. Applied Thermal Engineering, 220, 119, 876. https://doi.org/10.1016/j.applthermaleng.2024.119876.
[20] Singh R. & García E. (2025). Recycling-driven thermal-electrochemical analysis of end-of-life Li-ion cells. Sustainable Materials and Technologies, 35, e00567. https://doi.org/10.1016/j.susmat.2025.e00567.
[21] QC/T 743-2006, Lithium-ion batteries for electric vehicles [S].
[22] GB/T 31485-2015, Safety requirements and test methods for power batteries for electric vehicles [S].
[23] GB/T 31467.3-2015, Lithium-ion power battery packs and systems for electric vehicles, Part 3: Safety requirements and test methods [S].
[24] IEC 62660-2-2010, Secondary lithium-ion cells for the propulsion of electric road vehicles Part 2: Reliability and abuse testing [S].
[25] Huang M. & Zhang W. (2025). Graph-based thermal network models for large-scale battery storage systems. IEEE Transactions on Industrial Electronics, 72 (4), 3, 456-3, 465. https://doi.org/10.1109/TIE.2025.1234567.
[26] Zhang Y., Liu Q., & Patel R. (2024). Multiphysics modeling of lithium-ion batteries: Coupled electrochemical-thermal behavior at high C-rates. Journal of Power Sources, 592 (3), 123, 456. https://doi.org/10.1016/j.jpowsour.2024.123456.
[27] Lee S. & Chen W. (2025). Data-driven thermal-electrochemical coupling model for battery safety prediction. Applied Energy, 368, 234, 501. Preprint. arXiv. https://arxiv.org/abs/2402.XXXXX.
[28] Wang H., Zhang L., & Srinivasan V. (2024). A reduced-order electro-thermal model for fast simulation of large battery packs. IEEE Transactions on Transportation Electrification, 10 (4), 789-801. https://doi.org/10.1109/TTE.2024.987654.
[29] García R., Müller F., & Tanaka K. (2025). Phase change materials in Li-ion batteries: Integrated thermal-electrochemical modeling. Electrochimica Acta, 480, 129, 876. https://doi.org/10.1016/j.electacta.2025.129876.
[30] Kim J., Park S., & Gupta A. (2025). Fast charging optimization with electro-thermal-aging coupled models. TechRxiv. Preprint. https://doi.org/10.36227/techrxiv.987654321.
[31] Martinez A., Zhou T., & Kim D. (2025). Real-time thermal-electrochemical coupling in fast-charging Li-ion batteries: A control-oriented approach. Energy Storage Materials, 25, 100, 888. https://doi.org/10.1016/j.ensm.2025.100888.
[32] Kato T. & Singh M. (2024). Heterogeneous thermal-electrochemical modeling of solid-state batteries with 3D microstructure analysis. Advanced Energy Materials, 14 (45), 1, 234. https://doi.org/10.1002/aenm.202401234.
[33] Zhou X., Li H., & Wang Y. (2024). Machine learning-augmented multi-physics models for battery thermal runaway prediction. In 2024 IEEE Energy Conversion Congress and Exposition (ECCE) (pp. 1-8). IEEE. https://doi.org/10.1109/ECCE.2024.9999999.
[34] Nguyen V., Smith J., & Zhang Q. (2025). A transient electro-thermal model for lithium-sulfur batteries under low-temperature operation. Journal of the Electrochemical Society, 172 (2), 020543. https://doi.org/10.1149/1945-7111/ad3333.
[35] Zhang R. & Li Y. (2025). Coupled electrochemical-thermal simulation of battery modules using graph neural networks. SSRN. Preprint. https://ssrn.com/abstract=987654.

Journal of Physics: Conference Series **3031** (2025) 012007 doi:10.1088/1742-6596/3031/1/012007

Studies on the hybrid model and flow characteristics in vanadium redox flow battery

Song Li*, Yali Zhang, Jiahong Xun

School of Electrical Engineering and Automation, Hefei University of Technology, Hefei 230009, China

*Corresponding author's e-mail: lisong_1999@163.com

Abstract. To improve the operating efficiency of vanadium redox flow battery (VRB), this paper considers the effects of electrolyte flow rate and charge-discharge current on the VRB operation state. By coupling the equivalent loss model, the electrochemical model, and the pump loss model to establish a hybrid model for VRB. Through simulation analysis of the established model, obtain the influence rules of electrolyte flow rate and current change on VRB operating efficiency, thereby obtaining the optimal flow rate corresponding to currents of different sizes that maximizes operating efficiency. Real-time control of the electrolyte flow rate based on the size of the current can improve operating efficiency.

1. Introduction

In recent years, with the development of energy technologies, VRB has been an important solution for large-scale energy storage systems due to its excellent scalability, long lifespan, and superior safety features [1-2]. To further improve the efficiency of VRB, many scholars have conducted extensive research.

An accurate numerical simulation model is the foundation of system simulation calculations, so constructing a comprehensive and accurate VRB model is the primary content of this paper. Literature studies the knowledge related to battery models, as well as the system architecture of electrochemical energy storage standards[3-4]. Literature investigates the impact of relevant parameters of vanadium battery pipelines and electrodes on pump losses in single-stack VRB [5-6]. Literature establishes a shunt current equivalent model and studies its impact on coulombic efficiency and energy efficiency. It simply represents the battery with a DC power source, ignoring the complex variations and coupling relationships within the battery [7-8].

Based on the above, this paper couples the pump loss model, the equivalent loss model, and the electrochemical model of VRB. The hybrid model can accurately demonstrate the input-output characteristics of VRB. Through simulation analysis of the established models, the influence patterns of parameters such as electrolyte flow rate and charge-discharge current on the system efficiency of the VRB are obtained. By selecting the corresponding optimal flow rates for different charge-discharge currents under which VRB operates, the system efficiency of VRB can be effectively improved.

2. Structure and principle of VRB

As shown in Figure 1, the VRB system is primarily composed of a stack, two independent positive/negative electrolyte tanks, circulation pumps, and pipes. The vanadium electrolyte is stored in the storage tanks and within the stack. The stack consists of multiple series-connected individual cells, which mainly include ion exchange membranes, electrodes, bipolar plates, flow frames, and sealing components. Within each cell, the positive and negative electrolytes are separated by an ion exchange membrane. Vanadium electrolytes exist in four forms of vanadium ions: VO^{2+} and VO_2^+ for the positive electrode electrolyte, and V^{2+} and V^{3+} for the negative electrode electrolyte. Additionally, the stack is externally connected to loads or power sources, with circulation pumps and pipelines connecting the stack to the storage tanks.

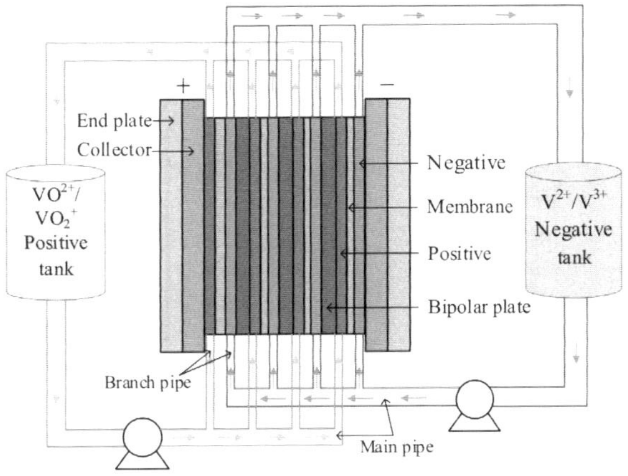

Figure 1. The structure and working principle of VRB.

During the charging process of a VRB using an external power source, vanadium ions undergo redox reactions at the electrodes within the cell stack. At the positive electrode, VO^{2+} continuously loses electrons and gradually oxidizes to VO_2^+. At the negative electrode, V^{3+} gains electrons and is progressively reduced to V^{2+}. Under the pressure exerted by the circulation pump, the vanadium electrolyte from the storage tanks flows into the cell stack, evenly passing through each cell, allowing the reaction to continue.

When the VRB discharges under load, VO_2^+ at the positive electrode gains electrons and is reduced back to VO^{2+}, while simultaneously, V^{2+} at the negative electrode continuously loses electrons and oxidizes to V^{3+}. The ion-exchange membrane only allows hydrogen ions to pass through in order to balance the internal charge of the battery. During charging, the concentrations of V^{3+} and VO^{2+} ions gradually decrease, whereas those of V^{2+} and VO_2^+ increase; this trend reverses during discharging.

Positive electrode reaction:

$$VO^{2+} + H_2O - e \underset{\text{charge}}{\overset{\text{discharge}}{\rightleftharpoons}} VO_2^+ + 2H^+ \tag{1}$$

Negative electrode reaction:

$$V^{3+} + e \underset{\text{charge}}{\overset{\text{discharge}}{\rightleftharpoons}} V^{2+} \tag{2}$$

3. Hybrid model of VRB

This paper establishes a hybrid model based on an equivalent loss model that takes into account electrochemical and fluid mechanics factors. It ignores the effects brought by temperature factors, differences in electrolyte distribution, and differences between battery stacks to simplify calculations.

3.1. Electrochemical model

Electrochemical model considering the influence factors of vanadium ions in storage tanks and stacks, quantifying the transfer of vanadium ions through membranes, combining charge-discharge redox reactions, crossover self-discharge reactions, and Fick's law to establish mass balance equations. It is known that the stack voltage is related to the number of individual cells and the equilibrium potential by combining a mass balance equation with an electrical circuit model. Dynamic differential equations for vanadium ion concentrations in the storage tank and stack are shown as Equations (3) - (6).

$$\begin{cases} \frac{V_t}{2}\frac{dC_{t2}}{dt} = Q(C_{s2} - C_{t2}) \\ \frac{V_s}{2}\frac{dC_{s2}}{dt} = Q(C_{t2} - C_{s2}) \\ -N\frac{S}{d}(k_2C_{s2} - 2k_5C_{s5} - k_4C_{s4}) \pm N\frac{I_d}{zF} \end{cases} \tag{3}$$

$$\begin{cases} \frac{V_t}{2}\frac{dC_{t3}}{dt} = Q(C_{s3} - C_{t3}) \\ \frac{V_s}{2}\frac{dC_{s3}}{dt} = Q(C_{t3} - C_{s3}) \\ -N\frac{S}{d}(k_3C_{s3} + 3k_5C_{s5} + 2k_4C_{s4}) \mp N\frac{I_d}{zF} \end{cases} \tag{4}$$

$$\begin{cases} \frac{V_t}{2}\frac{dC_{t4}}{dt} = Q(C_{s4} - C_{t4}) \\ \frac{V_s}{2}\frac{dC_{s4}}{dt} = Q(C_{t4} - C_{s4}) \\ -N\frac{S}{d}(k_4C_{s4} + 3k_2C_{s2} + 2k_3C_{s3}) \mp N\frac{I_d}{zF} \end{cases} \tag{5}$$

$$\begin{cases} \frac{V_t}{2}\frac{dC_{t5}}{dt} = Q(C_{s5} - C_{t5}) \\ \frac{V_s}{2}\frac{dC_{s5}}{dt} = Q(C_{t5} - C_{s5}) \\ -N\frac{S}{d}(k_5C_{s5} - 2k_2C_{s2} - k_3C_{s3}) \pm N\frac{I_d}{zF} \end{cases} \tag{6}$$

In the formulation, V_s and V_t represent the electrolyte volumes in the stack and the storage tank, respectively. C_{si} and C_{ti} (i=2,3,4,5) are the concentrations of i-valent vanadium ions. k_i is the membrane diffusion coefficient of i-valent vanadium ions. F is the Faraday constant. z is the electronic transfer coefficient. d is the thickness of the ion-exchange membrane. S is the area of the ion-exchange membrane. Q represents the flow rate of electrolyte into the stack. I_d indicates the magnitude of the charge-discharge current. N is the number of batteries. Based on the concentration of vanadium ions, using the Nernst equation can yield the open-circuit cell voltage U_{cell} for each cell. Assuming consistent properties among all cells, the stack voltage U_s can be calculated based on the individual cell voltages.

$$U_s = N \cdot U_{cell} = N\left[U_{eq} + \frac{RT}{zF}\ln\left(\frac{C_{s2} \times C_{s5}}{C_{s3} \times C_{s4}}\right)\right] \tag{7}$$

In the Nernst equation, U_{eq} represents the standard electrode potential of the positive and negative electrodes in a VRB undergoing an electrochemical reaction at 298 K, which is 1.259 V. R is the universal gas constant. T characterizes the effect of temperature on battery operation.

3.2. Fluid mechanics model

In addition to the electrochemical reactions within the stack, it is also necessary to consider the impact of electrolyte flow rate on the battery due to the circulation pump and the structure of the stack in a VRB. When the electrolyte flows through pipes, stack electrodes, and stack flow channels, it can generate a pressure drop, resulting in pump power loss. Pump loss P_{pump} mainly includes the flow pressure drop loss $\Delta P_{electrode}$ caused by electrolyte flowing through the internal electrodes, pipe friction loss ΔP_{pipe} caused by electrolyte flowing through external pipelines, and flow pressure drop loss ΔP_{stack} caused by electrolyte flowing through stack channels. Based on Darcy's law as shown in Equation (8), the flow pressure drop within the electrode can be calculated, which is related to the length L_e of the electrode, cross-sectional area A_{cs}, permeability K_e, and viscosity μ of the electrolyte. Since this paper does not discuss the specific layout method of the pipeline, only the straight pipe section friction resistance pressure drops, as shown in Equation (9), is considered for the calculation of the pipeline part pressure drop. L_n and D_h are the length and hydraulic diameter of the pipeline, respectively, and f_d is the resistance coefficient. The internal flow channel of the stack is relatively complex, and the literature simplifies the calculation method of flow resistance using finite element methods as shown in Equation (10), whose data is cited here, R_e [10].

$$\Delta P_{\text{electrode}} = \frac{\mu L_e Q}{K_e A_{cs}} \tag{8}$$

$$\Delta P_{\text{pipe}} = \frac{f_d L_n \rho v^2}{2 D_h} \tag{9}$$

$$\Delta P_{\text{stack}} = Q \cdot R_e \tag{10}$$

$$I_{\text{pump}} = \frac{P_{\text{pump}}}{U_s} = \frac{Q \sum_j \Delta P_j}{\eta U_s} \tag{11}$$

In the formula, η represents pump efficiency, and the subscript j in ΔP indicates the flow path of the electrolyte through the pipe.

3.3. Hybrid model

As shown in Figure 2, the VRB hybrid model is divided into three models: the equivalent loss circuit model, the electrochemical model, and the fluid mechanics model.

Literature proposes a VRB equivalent loss model with a simple structure and clear physical meaning[9]. The internal resistance includes the ohmic internal resistance R_b and the reaction internal resistance R_a. The fixed parasitic resistance R_f represents the loss resistance caused by the system controller and stack shunt current. I_p represents the variable pump loss current. The electrode capacitance C_e reflects the transient performance of the battery during charging and discharging. U_s represents the stack voltage (open-circuit voltage, OCV). U_d is the terminal voltage of the battery. I_s represents the current flowing through the stack. I_e represents the capacitor current. The reference directions of various currents are shown in Figure 2.

In the electrochemical model, the input includes the charge-discharge current size I_d, VRB system parameters, electrolyte flow rate Q, and initial concentrations of ions in different valence states. Through the dynamic differential equations of electrochemical reactions (3) - (6), the real-time concentrations of vanadium ions at various valence states are obtained. The SOC value is calculated based on the concentration of vanadium ions in the storage tank. The open-circuit voltage U_{cell} of individual cells is obtained through the Nernst Equation (7) using the concentration of vanadium ions in the stack. Given that the properties of each cell are consistent, calculate the stack voltage U_s based on the number of cells.

In the fluid mechanics model, based on the flow rate of the circulation pump and the basic characteristics of the VRB hydraulic loop, the electrode pressure drop $\Delta P_{\text{electrode}}$, the stack pressure drop ΔP_{stack}, and the pipe pressure drop ΔP_{pipe} caused can be calculated. Then it can obtain the total pressure drop ΔP to get the pump loss power P_{pump} according to the flow rate and pump efficiency. Considering the terminal voltage feedback from the all-vanadium redox flow battery system, it can obtain the pump loss current I_{pump}. Using I_{pump} as a controlled variable, the current source in the equivalent circuit. Other losses in the system are equivalent to resistances, and the parameter values in the equivalent circuit can be calculated. Combining the controlled voltage source U_s and the controlled variable current sources I_{pump}, the equivalent operating terminal voltage U_d of the VRB can be estimated.

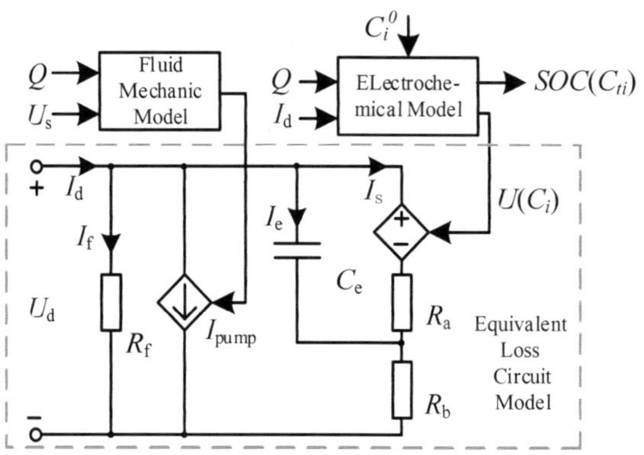

Figure 2. Hybrid model diagram of VRB.

4. Efficiency analysis of VRB

The system efficiency of VRB η_s reflects the input-output capability of the system. The calculation formula is as follows:

$$\eta_s = \frac{\int_0^{t_d} P_s dt - \int_0^{t_d} P_{pump} dt}{\int_0^{t_d} P_s dt + \int_0^{t_d} P_{pump} dt} \tag{12}$$

In the formula, P_s is the rated power of the battery.

According to the working principle and model analysis of VRB, when the flow rate Q is constant and the current I increases, the flow velocity of ions in vanadium electrolyte will be lower than the reaction velocity of ions, resulting in energy loss and thus reducing system efficiency. When the current I decreases, the flow velocity of ions in the vanadium electrolyte will be higher than the reaction velocity of ions, causing greater mechanical losses and thus reducing system efficiency.

In a constant current condition, there is only one control variable, the electrolyte flow rate Q. The power of the fuel cell stack and pump loss are affected by the flow rate. When the flow rate Q is too high, it increases mechanical losses in the battery pump, thereby reducing the efficiency of the battery system. On the other hand, when the flow rate Q is too low, it cannot deliver sufficient electrolyte for the reaction, resulting in reduced battery power and lower battery efficiency. Therefore, there must be an optimal flow rate that maximizes the efficiency of the battery during operation, thus minimizing the system efficiency."

5. Simulation studies

5.1. Charge-discharge state simulation

The simulation uses the GEC-VRB-5kW/20 kW·h VRB battery, which consists of a 5 kW stack and 1.5 mol/L vanadium electrolyte. The stack is composed of 40 individual cells. Table 1 lists the specifications of the VRB and the parameters used in this simulation.

Table 1. Specifications of the VRB and parameters in the simulation.

Parameters	Symbol	Value
Rated power	P_s (kW)	5
Rated voltage	U_s (V)	48
Rated current	I_d (A)	100
Charge/discharge time	T_d (h)	4
Number of cells	N	40

Tank volume	V_t (L)	1600
Stack volume	V_s (L)	160
Membrane thickness	d (dm)	2.54×10^{-3}
Membrane area	S (dm^2)	10
transfer electrons	Z	1
Faraday constant	F	96485
Vanadium concentration	C (mol L^{-1})	1.5
Diffusion coefficient of V^{2+}	k_2 (dm^2 s^{-1})	8.768×10^{-10}
Diffusion coefficient of V^{3+}	k_3 (dm^2 s^{-1})	3.222×10^{-10}
Diffusion coefficient of VO^{2+}	k_4 (dm^2 s^{-1})	6.825×10^{-10}
Diffusion coefficient of VO_2^+	k_5 (dm^2 s^{-1})	5.897×10^{-10}
Internal resistance	R_a (Ω)	0.040
Reactive resistance	R_b (Ω)	0.026
Fixed loss resistance	R_f (Ω)	13.94
Dynamic response capacitor	C_e (F)	0.154
Electrolyte density	ρ (g L^{-1})	1354
Electrolyte viscosity	μ (mPa s^{-1})	4.928
Pump efficiency	η	0.8
Ratio of main pipe length to cross-sectional area	L_m/A_m (dm^{-1})	460
Ratio of branch pipe length to cross-sectional area	L_n/A_n (dm^{-1})	4600
Flow resistance	R_e (Pa L^{-1})	14186.843
Standard electrode potential	U_{eq} (V)	1.259

In order to verify the accuracy of the VRB hybrid model, after building the mentioned hybrid model in MATLAB/Simulink and setting the parameters as described above, a simulation analysis is conducted. A constant current of 100A is applied for charging and discharging simulations on the battery system. The concentrations of V^{2+}, V^{3+}, VO^{2+}, and VO_2^+ 4 vanadium ions are set at 0.15, 1.35, 1.35, and 0.15 mol/L, respectively. The flow rate Q is kept constant at 0.3 L/s.

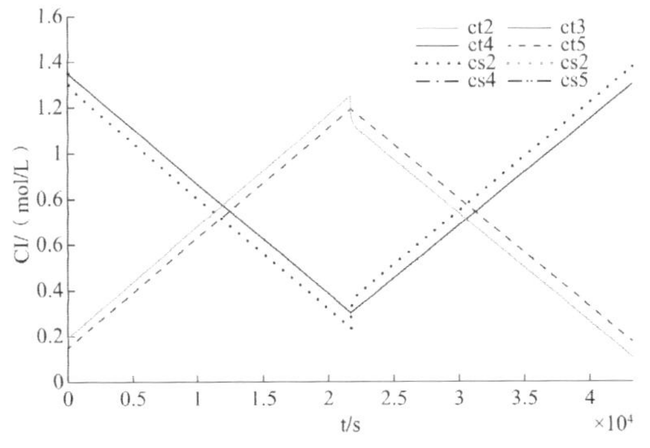

(a)

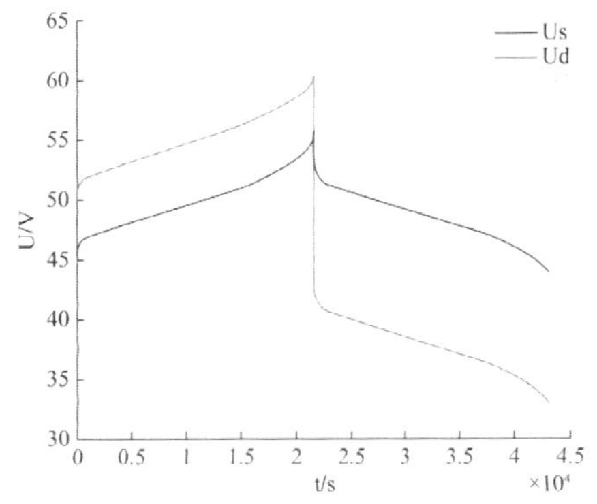

(b)

Journal of Physics: Conference Series **3031** (2025) 012007 doi:10.1088/1742-6596/3031/1/012007

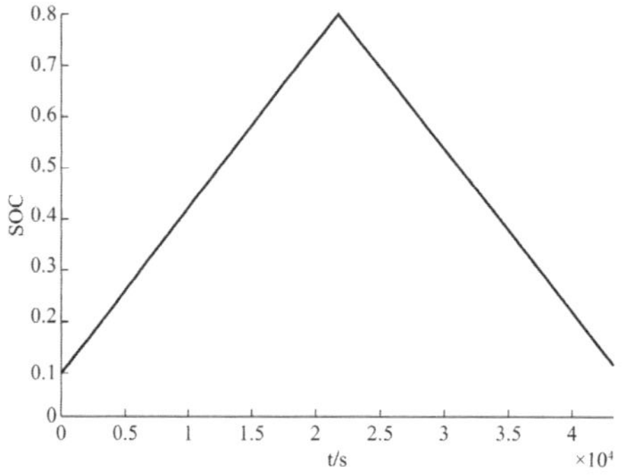

(c)

Figure 3. Variations of vanadium ion concentration, stack and open circuit voltage, and SOC of the battery during charging and discharging.

During the charging process, as shown in Figure 3(a), $C_{s2} > C_{t2}$ and $C_{s5} > C_{t5}$ while $C_{s3} < C_{t3}$ and $C_{s4} < C_{t4}$. In contrast, during the discharging process, $C_{s2} < C_{t2}$ and $C_{s5} < C_{t5}$, whereas $C_{s3} > C_{t3}$ and $C_{s4} > C_{t4}$. This is due to the oxidation-reduction reactions occurring during charge-discharge cycles. The results indicate that this model generally aligns with the fundamental principles of electrochemical reactions.

Figure 3(b) shows the variation of battery terminal voltage and stack voltage during charging and discharging of VRB at a constant current of 100A. It can be seen from the figure that $U_d > U_s$ during the charging process, and $U_d < U_s$ during the discharging process. The difference is caused by the mechanical loss of all-vanadium redox flow batteries. When switching between charging and discharging, the direction of the current changes, resulting in corresponding changes in the battery's losses.

Figure 3(c) shows the variation of SOC with time during charging and discharging processes for VRB. It can be seen that SOC exhibits a linear change trend, and $0.1 < SOC < 0.9$ within this range, overcharging and deep discharge phenomena will not occur in the battery. According to the above simulation results, it is known that the operation result of the model conforms to the working principle of the battery.

5.2. The impact of electrolyte flow rate on batteries

By simulating different flow rate values Q under various current values, the system efficiency η_s of the battery is obtained for different conditions. For different current values, during charging and discharging states, the flow rate that maximizes the system efficiency η_s is considered the optimal flow rate.

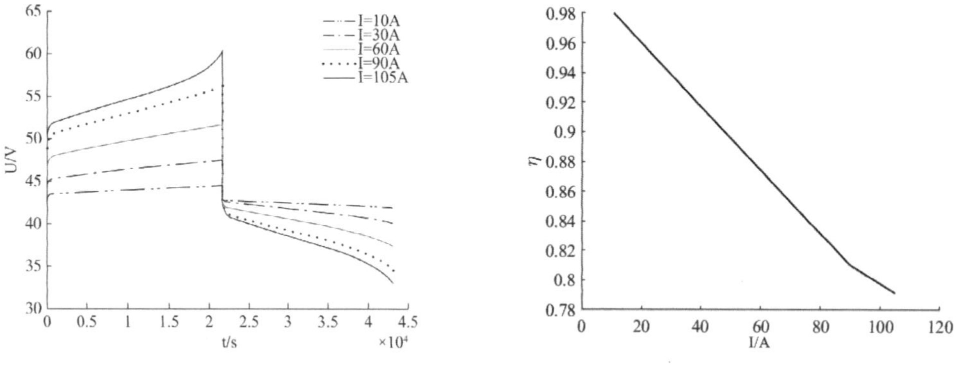

(a) (b)

Figure 4. Variation curves of VRB terminal voltage and system efficiency under different charge-discharge currents.

Figure 4(a) shows the voltage variation curves of the battery stack when charging and discharging at constant currents of 10 A, 30 A, 60 A, 90 A, and 100 A with an electrolyte flow rate of 0.3 L/s. The system efficiency variation curve is shown in Figure 4(b). From the graph, it can be seen that as the current increases, the charging voltage becomes higher and the discharging voltage becomes lower. The system efficiency decreases from 97.8% to 78.9%, which verifies the previous inference: if the current increases, the flow speed of ions in the vanadium electrolyte will be lower than their reaction speed, leading to energy loss when the flow rate Q is constant and thus reducing the system efficiency.

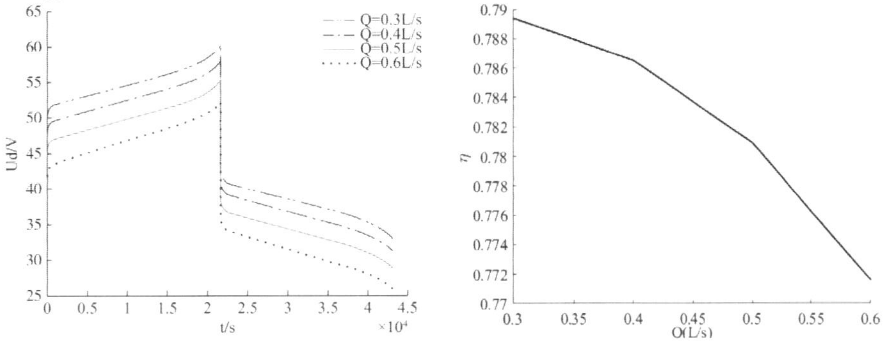

(a) (b)

Figure 5. Variation curves of VRB terminal voltage and system efficiency under different flow rates.

Figure 5 shows the battery terminal voltage and system efficiency curves for different flow rates of 0.3 L/s, 0.4 L/s, 0.5 L/s, and 0.6 L/s when charging and discharging the battery at a constant current of 100 A. From the graph, it can be seen that as the flow rate increases, the charge-discharge voltage decreases. The system efficiency drops from 78.9% to 77.1%, which proves that when the flow rate Q is too high, it leads to increased mechanical losses in the battery pump and thus reduces the efficiency of the battery system. On the other hand, when the flow rate Q is too low, there isn't enough electrolyte transported for reactions, leading to a decrease in battery power and thus reducing battery efficiency. If we ignore the size of the current and only increase the flow rate, not only will it fail to improve the battery's efficiency, but also increase system losses and reduce overall efficiency. Therefore, during the operation of the battery, choosing an appropriate flow rate based on the current size can better enhance the operational efficiency of the battery.

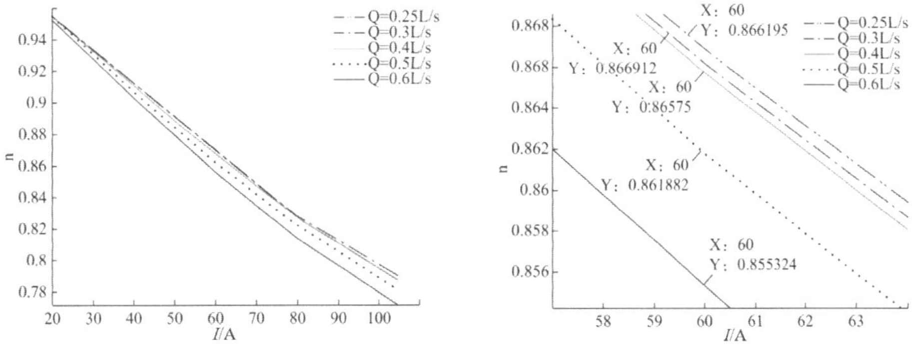

(a) (b)

Figure 6. The relation curve between system efficiency and flow rate, and its close-up.

Figure 6 shows the variation curves of battery system efficiency when the current belongs to [20 A – 100 A] and Q is at different flow rates: 0.25 L/s, 0.3 L/s, 0.4 L/s, 0.5 L/s, and 0.6 L/s. From the graph, it can be seen that as current increases, the improvement in battery efficiency due to changes in flow rate also increases. To facilitate numerical analysis, Figure 6(b) is enlarged to obtain Figure 6(a).

Table 2. Ion concentration corresponding to the initial SOC.

Current(A)	Optimal flow rate(L/s)	Current(A)	Optimal flow rate(L/s)
0-10	0.41	50-60	0.31
10-20	0.39	60-70	0.29
20-30	0.37	70-80	0.27
30-40	0.35	80-90	0.25
40-50	0.33	90-100	0.23

According to Figure 6(b), when the current is 60 A, setting $Q = 0.3$ L/s results in a system efficiency increase of 1.03% compared to setting $Q = 0.6$ L/s. When I is constant and Q is increased, the system efficiency first increases and then decreases. Similarly, each current has a corresponding optimal flow rate that maximizes the corresponding system efficiency. The optimal flow rate falls within the range of approximately 0.2 L/s - 0.4 L/s.

Based on the general trend shown in Figure 6(b), the flow rate is divided equally by 0.02 L/s within the range of 0.23 L/s - 0.41 L/s for simulation to determine the optimal flow rate values corresponding to each current. The specific values are listed in Table 2.

6. Conclusion

Based on the results and discussions presented above, the conclusions are as below:

(1) Simulation validation and analysis of the model demonstrate that the established hybrid model meets modeling requirements and considers electrical relationships, mass transfer processes, as well as the correlation between flow rates and pump losses, thereby improving the accuracy of the model.

(2) By analyzing the effects of charge-discharge currents and electrolyte flow rates on system efficiency, and studying their corresponding influence patterns, it is concluded that adopting variable flow rate control during VRB operation, selecting optimal electrolyte flow rates based on different current magnitudes, can effectively improve the operational efficiency of VRB. When the current is 60 A, the system efficiency increases by 1.04% by increasing the flow rate from 0.3 L/s to 0.6 L/s.

(3) By equally dividing the flow and simulating each segment, we obtained the optimal velocity ranges corresponding to different current segments. This will help with future VRB grid integration and reduce the overall energy consumption of the system.

References

[1] CHEN Haisheng, LI Hong, XU Yujie, et al. Research progress on energy storage technologies of China in 2023[J]. Energy Storage Science and Technology, 2024, 13(5): 1359-1397.

[2] Khan K M, Raza M, Shahbaz M, et al. Recent advancement in energy storage technologies and their applications[J]. Journal of Energy Storage,2024,92112112-.

[3] LIU Lu, NIU Meng, LI Jianlin, et al. Research on the status quo and system architecture of electrochemical energy storage system standards[J]. Electric Power Construction, 2020, 41(4): 63-72.

[4] LI Jianlin, GUO Zhaodong, ZENG Wei, et al. Modeling and simulation analysis of lithium battery energy storage for frequency regulation[J]. Power System Protection and Control, 2022, 50(13): 33-42.

[5] Prasanna M M, Jayanti S Effect of electrode thickness and compression on the performance of a flow-through mode vanadium redox flow battery cell[J]. Journal of Power Sources,2025,631236203-236203.

[6] Aumber A, Saleem A, Arjun B, et al. Effect of electrode porosity on the charge transfer in vanadium redox flow battery[J]. Journal of Power Sources, 2021,488.

[7] LI Minghua, WANG Baoguo, FAN Yongsheng. Measurement and simulation of bypass current in all-vanadium liquid current batteries[J]. Power Supply Technology, 2020, 44(3): 402-407.

[8] YU Shuyuan, YE Qiang. Liquid flow battery pipeline optimization method based on efficient compact design[J]. Power Supply Technology, 2018, 42(11): 1694-1697.

[9] CHAHWAN J, ABBEY C, JOOS G. VRB modelling for the study of output terminal voltages, internal losses and performance[C]//Electrical Power Conference, IEEE, 2007.

[10] BLANC C, RUFER A. Multiphysics and energetic modeling of a vanadium redox flow battery[C]. IEEE International Conference on Sustainable Energy Technologies, 2008.

Journal of Physics: Conference Series **3031** (2025) 012008 doi:10.1088/1742-6596/3031/1/012008

Life-cycle operational management for electrochemical energy storage

Qi Zhang [1], Luye Yang [2*], Zhengrun Wu [3], Xiaojing Li [2], Qihang Gong [2]

[1]China Three Gorges Renewables (Group) Co., Ltd., Beijing, China
[2]PKU-Changsha Institute for Computing and Digital Economy, Changsha, China
[3]Peking University, Beijing, China

[*]Corresponding author's e-mail: yangluye@icode.pku.edu.cn

Abstract. Over the past several years, electrochemical energy storage (EES) technology has gradually moved toward commercial operation due to its high efficiency. However, the scientific evaluation of its long-term value still poses challenges. Battery life (including economic and physical life) is a critical factor in the economic assessment and operation of EES. Due to the complexity of battery life, the discrepancy between economic and physical lives may lead to biases in the economic evaluation of EES systems. In this study, a comprehensive full life cycle decision-making model is constructed to provide a scientific basis for the planning, operation, and decommissioning decisions of energy storage systems (EES). The model aims to systematically evaluate the economic benefits of EES over its life cycle and determine the marginal degradation cost (MDC), which is analyzed through simulation, and found that the results show that for distributed/residential EES systems, the endpoint of economic life may precede that of physical life.

1. Introduction

With the global energy transition, integrating volatile renewable energy into the grid challenges power system stability and efficiency. EES is a key solution for balancing supply and demand, enhancing energy efficiency, and supporting clean energy transition [1,2]. EES stores excess renewable energy and releases it during peak demand, optimizing grid performance. EES in electricity markets is equally vital, providing essential services such as energy arbitrage [3], peak shaving, and voltage stabilization [4]. Recent advancements have expanded EES applications to microgrids, electric vehicle charging, and emergency backup power.

EES is crucial for low-carbon energy systems, but its performance degrades over time. Degradation is categorized into cycling degradation, influenced by cycle count, DOD (depth of discharge), charge/discharge rate, SOC (state of charge), management system efficiency [5,6], and calendar degradation, which depends on time and storage conditions. Degradation impacts EES performance, reliability, and economic returns.

Accurate battery degradation evaluation is essential for optimizing short-term operations and long-term investments. To accurately estimate battery state of health (SOH) and predict its cycle life, researchers have proposed and established multiple distinct methodologies. Among these, Physics-based models (PBMs) [7], rooted in electrochemical principles and internal battery processes, have been extensively researched. Equivalent circuit models (ECMs) [8], which represent battery behavior using simplified electrical circuits, are widely adopted for their computational efficiency. Advances in machine learning models (MLs) [9,10] have enabled high-accuracy predictions through data-driven methods, bypassing the need for detailed internal battery knowledge. Meanwhile, empirical models (EMs) [11], built on experimental data and statistical relationships, remain practical for applications where comprehensive physical details are unavailable. These models integrate degradation into EES

Journal of Physics: Conference Series **3031** (2025) 012008 doi:10.1088/1742-6596/3031/1/012008

operation and maintenance decisions. Cycling degradation is modeled using cycle count, temperature, and DOD, while calendar degradation considers time, temperature, and SOC. Managing degradation costs is critical for maximizing EES benefits. He et al. [3,12] incorporated degradation costs into optimization models, enhancing power system and EES decision-making [13]. However, lifecycle management and cost-benefit analysis of EES remain understudied.

This paper proposes a lifecycle decision model for EES, focusing on charge/discharge scheduling and battery health management to extend cycle life and optimize marginal degradation cost (MDC). It explores the impact of battery economic and physical life on EES and analyzes cycling and impacts of calendar degradation on energy arbitrage profitability based on economic end-of-life (EOL) criteria. The findings provide valuable insights for EES planning and operation.

Table 1. Indices, Variables, and Functions.

Indices, Variables, and Functions			
x	Optimal MDC use costs	E_h^e	The electricity market price
ϕ	EES life cycle revenue	θ	DOD
g_t	The highest short-period gain within the period t	$N_{t,\theta}$	cycle numbers at DOD during $[t, t + \Delta t]$
d_t	EES degradation within the period t	P_h^{dis}	EES discharge outputs
δ_t	discounting factor within the period t	P_h^{cha}	EES charging outputs
D	100%-depth of discharge throughput limit.	$E^{\max}$	Energy capacity of EES within the period t
T	EES life-time length (depending on its d_t and D)	$\mathbf{F}(s_t)$	The feasible operating range of EES depends on its SOH within the period t.
q	Calendar degradation rate	e_h	SOC within the period h
$\boldsymbol{P}_t$	Charge/discharge schedules within the period t	σ_t	Charging/discharging efficiency within the period t
$f_t(\boldsymbol{P}_t)$	Short-term revenue within the period t	Γ	The self-discharge rate
$2E^{\max}$	The energy capacity of one cycle (both charging and discharging)	$P^{\max}$	The power capacity at time t

2. Methods

2.1. EES life-cycle decision framework

To emulate EES's operating decisions and assess its cash flows over its life cycle, we have developed a life-cycle decision analysis framework for modelling decision-making behaviour during the operation of an EES. (excluding initial capital costs during operation). MDC is introduced to bridge short- and long-term decisions, representing the battery's future utilization value and acting as the short-term marginal operational cost (opportunity cost). Using forecast data, the optimal MDC is determined by deriving optimal short-term operational strategies to maximize life-cycle revenue under battery degradation constraints. The mathematical formulas are as follows, with specific parameters and variables denoted as shown in Table 1:

$$\phi\, max = \max_x \phi = \max_x \textstyle\sum_{t \le T} \delta_t g_t(x) \tag{1}$$

$$\text{s.t.} \textstyle\sum_{t \le T} d_t(x) \le D \tag{2}$$

$$d_t(x) \ge q \tag{3}$$

$$g_t(x) = \max_{\boldsymbol{P}_t \in \boldsymbol{F}(s_t)} f_t(\boldsymbol{P}_t) \tag{4}$$

$$s.t.\, \boldsymbol{P}_t = \mathbf{0} \; or \; \frac{\partial g_t(x)}{\partial d_t(x)} = \frac{x}{\delta_t} \tag{5}$$

As outlined in Formulas (1) – (3), the EES operator maximizes the project's long-term net revenues under EES degradation constraints, requiring the operator to identify the optimal life-cycle MDC. (1) Summing the net revenues generated across various medium- and short-term simulation intervals provides the life-cycle net benefit. (2) Constrains total energy throughput by the physical EOL criterion, with the MDC equating to the Lagrange multiplier of (2). (3) Accounts for calendar degradation, reflecting time-dependent EES deterioration.

Within the mid-term planning period (typically one month to one year), the EES operator adjusts the MDC by applying a discount factor to the life-cycle MDC, incorporating long-term insights and time-based profitability preferences. This adjustment promotes higher EES utilization in early years. The revised MDC is then embedded in short-term planning to align operational decisions with life-cycle benefit maximization.

For the short term, which is generally defined as a day. The EES operators maximize daily earnings by determining the optimal short-term output scenario, whether on an hourly or within-hour time scale. In developing this scenario, the following key factors need to be considered: the SOH of the unit, the forecasted spot market price of electricity, and the mid-term update of the adjusted MDC, as shown in Formulas (4) and (5). Formula (5) (first-order Karush–Kuhn–Tucker condition) is considered to be the optimal necessary condition for the long-term maximization problem ((1) - (3)).

2.2. *The model of energy arbitrage in short-term*

The formulas used for the model of decision optimization in short-term are shown in (6) – (11):

$$g_t = \max_{p_h^{dis},\, p_h^{cha}} \sum\nolimits_{h\in[t,t+\Delta t]} E_h^e \left(p_h^{dis} - p_h^{cha}\right) - C_{fix} \tag{6}$$

$$s.t. \frac{\partial g_t}{\partial d_t} = \frac{x}{\delta_t} if\ d_t > q_t \tag{7}$$

$$d_t = \sum\nolimits_\theta 2E^{\max} N_{t,\theta}\, \theta^k + q_t \tag{8}$$

$$e_h = (1-\Gamma)e_{h-1} + P_h^{cha}\sigma_t(s_t) - P_h^{dis}/\sigma_t(s_t) \tag{9}$$

$$0 \leq P_h^{dis}, P_h^{cha} \leq P^{\max} \tag{10}$$

$$0 \leq e_h \leq E^{\max} \tag{11}$$

As shown in Formula (6), within the time interval [t, t + Δt], short-term (intraday) net return for each hour h is the remainder of the operating revenue after deducting the fixed O&M costs. In general, the period Δt for short-term scheduling is set to a period of one day. Formula (8) is to equate the cyclic degradation at different discharge depths to 100%-DOD degradation energy throughput. For EES, one to two charge/discharge cycles per day are usually required. Therefore, we can set the cycle number of periods = 1, and the DOD, for energy arbitrage. Formula (9) is used to calculate the SOC during EES charging/discharging. During operation, the EES system must comply with physical limitations, including constraints on power output and the power supply's SOC, which are expressed in Formulas (10) and (11), respectively.

2.3. *Parameter setting*

The California Independent System Operator (CAISO) 2018 non-spinning standby price and the day-ahead market LMP data are used in this case as a reference for the tariff and ancillary services prices. When the battery reaches a total of 2,000 complete charge and discharge cycles under conditions of 100% DOD or when the battery capacity drops to approximately 80% of the initial capacity, it is considered to have arrived at its physical end-of-life. The calendar degradation rate of the battery is set to 1% per year [14], with an initial charge/discharge efficiency of 95% and a self-discharge rate ρ = 0. The annual fixed O&M cost for utility-scale lithium EES projects is set at 9/kW-year, and the annual fixed O&M costs for utility and distributed/residential lithium EES are 9 and 27/kW-year, respectively.

3. Results

3.1. Physical and economic life

EES economic EOL is reached when operations no longer yield net positive economic benefits, necessitating decommissioning and re-purposing or recycling if it occurs before the physical EOL. This study focuses on a grid-connected EES in California performing energy arbitrage—By charging when electricity prices are relatively low and discharging when prices peak, the system facilitates the improved integration of renewable energy into the grid. Energy arbitrage is highlighted as a critical strategy for low-carbon energy systems.

Figure 1 depicts the cash flows resulting from optimal operational strategies for utility-scale and distributed EES, highlighting differences driven by fixed O&M costs. In both cases, net annual profits decline over time, primarily due to the battery's performance deterioration and the reduction in profit value over time from the perspective of owners/operators, with degradation having a more significant effect. When the fixed O&M cost is set at 9/kW-year for utility-scale EES systems. Figure 1(a) shows that the net profit of the EES diminishes annually, with both the physical and economic EOL occurring after five years. Figure 1(b) shows net annual profits turning negative by the fifth year, failing to cover fixed O&M costs. Consequently, the economic EOL occurs at year 4, preceding the physical EOL. By the fifth year, the EES may no longer yield net profits from energy arbitrage in electricity markets, despite remaining operational. This underscores the likelihood that economic EOL often precedes the manufacturer's technical EOL, posing a practical challenge for EES owners and operators.

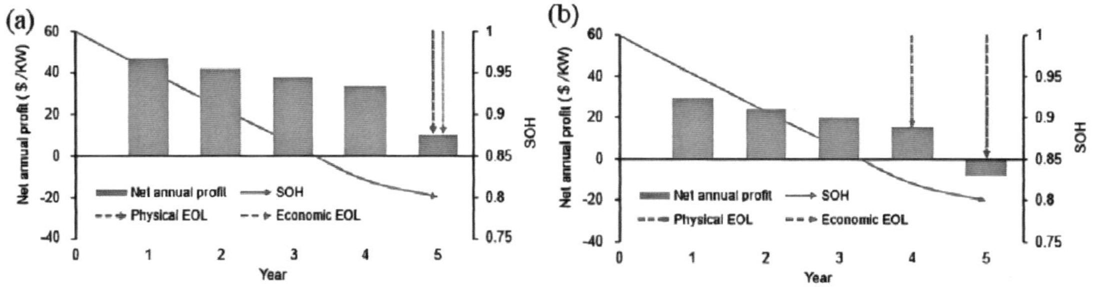

Figure 1. Cash flows of EES systems in California vary across application scenarios. (a) Fixed O&M costs for utility-scale large-scale energy storage systems are set at $9/kW per year. (b) Fixed O&M costs for distributed/ residential-scale energy storage systems are higher at $27/kW per year, energy arbitrage only.

3.2. Economic viability of ESS

Utilizing the model proposed, we evaluate the total revenue generated over the life cycle of EES and display the outcomes in Figure 2(a). The analysis breaks down revenue from energy arbitrage, ancillary services, and their combined total (life-cycle revenue), alongside the MDC. When the MDC rises from $0 - 10 per MWh-throughput, the life-cycle revenues are derived from energy arbitrage and ancillary services. Between $10-25/MWh-throughput, EES life-cycle revenue grows gradually, stabilize sat $25-60/MWh-throughput, followed by a steep decline beyond the $60/MWh−throughput, With an MDC setting of $45 per MWh-throughput, the EES system achieves the greatest level of net revenue over its entire life cycle. The primary source of EES revenue derives from arbitrage revenue at this time. This suggests that energy arbitrage using EES is economical when the MDC falls within the range of $10 to $60 per MWh of throughput. In other words, as the battery approaches the latter stages of its service life, the primary contributor to EES economic benefits becomes the revenue streams generated from energy arbitrage activities.

Journal of Physics: Conference Series **3031** (2025) 012008 doi:10.1088/1742-6596/3031/1/012008

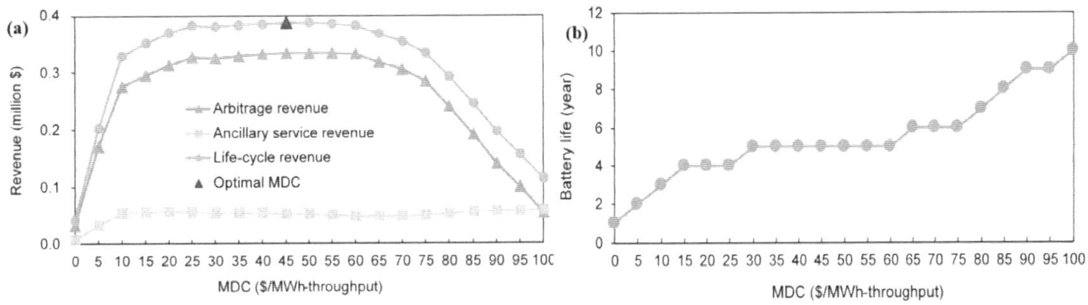

Figure 2. (a) Economic viability of ESS with different marginal degradation costs. (b) Battery life at different marginal degradation costs.

The benefits of energy arbitrage can cover the costs of battery degradation, so EES operators may be willing to accept higher MDC in markets with more volatility in electricity prices and more arbitrage opportunities. If there are limited opportunities for energy arbitrage or less volatility in electricity prices, the costs of battery degradation are likely to be a more significant portion of the cost, and operators must reduce the MDC to safeguard the batteries' long-term value. The EES operator needs to base their MDC on consideration of battery replacement costs and future market conditions to determine the ideal MDC for maximizing lifetime battery profitability. Here, the optimal MDC is $45/MWh-throughput.

3.3. MDC impact on battery lifetime

In this section, we explore the effect of MDCs on battery lifetime. As shown in Figure 2(b), the battery lifetime increases with the increase of MDC (i.e., 0 to $30/MWh-throughput), then stabilizes ($30 to $60/MWh-throughput), and finally experiences a sharp increase (from $60 to $100/MWh-throughput). MDC is an opportunity cost based on the value of the battery's future use, reflecting the cost per unit of battery degradation. When the MDC is relatively low (i.e., 0 ~ $60/MWh-throughput), EES operators may be more inclined to use the batteries frequently for arbitrage to pursue short-term profits, which will accelerate the degradation of the batteries and shorten their physical life. When a higher MDC (i.e., $60 ~ $100/MWh-throughput) implies a higher estimate of the battery's future value, it will motivate the operator to use the battery more cautiously to extend its lifetime.

4. Conclusion

This study establishes two critical concepts for EES systems: economic and physical EOL. Using the model presented in this study, we determine an optimal life-cycle profit and MDC while examining MDC's influence on battery lifespan. To accurately assess EES economic viability, annual fixed O&M costs are integrated into the life-cycle decision-making framework. Economic EOL occurs when revenues no longer cover these fixed costs, shifting from net profits to net losses. When analyzing lithium-ion EES performing energy arbitrage within the California electricity market as a practical example, utility-scale and distributed/residential systems reached economic EOL at 5 and 4 years, respectively, coinciding with a capacity reduction to 80% and a physical lifespan of 5 years. Economic EOL varies by EES scale, application, and degradation traits, necessitating detailed analysis of degradation patterns for specific business models or lifespan requirements. The reality that EES systems have an "economic end of life" may affect and change the planning and deployment strategy for the entire system. If EES owners fail to adequately consider this factor during the initial planning phase, likely, the power system will likely not be able to replace batteries promptly, resulting in additional economic losses.

Acknowledgments

This work has received financial support from the China Three Gorges Corporation, under the project "Economic Evaluation and Decision System Development for Electrochemical Energy Storage Systems" with the contract number: Three Gorges Renewables Contract [2023] No. 415.

References

[1] S. Wang, F. Li, G. Zhang, and C. Yin (2023) Analysis of energy storage demand for peak shaving and frequency regulation of power systems with high penetration of renewable. Energy, vol. 267, pp. 126586.
[2] X. Wei, S. Ban, X. Shi, P. Li, Y. Li, and S. Zhu et al (2023) Carbon and energy storage in salt caverns under the background of carbon neutralization in China. Energy, vol. 272, pp. 127120.
[3] Yang L Y, Cao J H, Cai B R, et al. (2021) Electrospun MOF/PAN Composite Separator with Superior Electrochemical Performances for High Energy Density Lithium Batteries. Electrochimica Acta, vol. 382(6058), pp.138346.
[4] Eladl A A, Fawzy S, Abd-Raboh E E, et al (2024) A comprehensive review on wind power spillage: Reasons, minimization techniques, real applications, challenges, and future trends. Electric Power Systems Research, 226, pp.109915.
[5] Júnior, C. A. R., Sanseverino, E. R., Gallo, P., Koch, D., Kotak, Y., Schweiger, H. G., & Zanin, H (2023). Towards a business model for second-life batteries: Barriers, opportunities, uncertainties, and technologies. Journal of Energy Chemistry, vol. 78, pp. 507-525.
[6] Mohammed, A. S., Atnaw, S. M., Salau, A. O., & Eneh, J. N (2023). Review of optimal sizing and power management strategies for fuel cell/battery/super capacitor hybrid electric vehicles. Energy Reports, vol. 9, pp. 2213-2228.
[7] M. Doyle, T. F. Fuller, and J. Newman (1993). Modeling of galvanostatic charge and discharge of the lithium/ polymer/ insertion cell. Journal of the Electrochemical Society, vol. 140, pp.1526.
[8] Che, Y., Hu, X., Lin, X., Guo, J., & Teodorescu, R (2023). Health prognostics for lithium-ion batteries: mechanisms, methods, and prospects. Energy & Environmental Science, vol. 16(2), pp.338-371.
[9] Ren, Z., & Du, C. (2023). A review of machine learning state-of-charge and state-of-health estimation algorithms for lithium-ion batteries. Energy Reports, vol. 9, pp. 2993-3021.
[10] K. Liu, X. Hu, Z. Wei, Y. Li, and Y. Jiang (2019). Modified Gaussian process regression models for cyclic capacity prediction of lithium-ion batteries. IEEE Transactions on Transportation Electrification, vol. 5, pp. 1225-1236.
[11] W. Vermeer, G. R. C. Mouli, and P. Bauer (2021) A comprehensive review on the characteristics and modelling of lithium-ion battery ageing. IEEE Transactions on Transportation Electrification, vol. 8, pp. 2205-2232.
[12] G. He, S. Kar, J. Mohammadi, P. Moutis, and J. F. Whitacre (2020). Power system dispatch with marginal degradation cost of battery storage. IEEE Transactions on Power Systems, vol. 36, pp. 3552-3562.
[13] X. Chen, Y. Yang, J. Wang, J. Song, and G. He (2023) Battery valuation and management for battery swap** station. Energy, vol. 279, pp. 128120.
[14] Kim, N., Kim, Y., Sung, J., & Cho, J. (2023). Issues impeding the commercialization of laboratory innovations for energy-dense Si-containing lithium-ion batteries. Nature Energy, vol. 8(9), pp. 921-933.

Journal of Physics: Conference Series **3031** (2025) 012009 doi:10.1088/1742-6596/3031/1/012009

Optimization of startup capability for the turbine bypass system in a 100MW tower concentrated solar power plant

Xinxin Xu*, Yanpu Wang, Junfeng Zhang, Yabing Niu

Department of New Energy Project, Powerchina Northwest Engineering Corporation Limited, Xi'an, China

*Corresponding author's e-mail: xuxinx@nwh.cn

Abstract. This paper addresses the common issue of slow startup speed in domestic concentrated solar power (CSP) plants by analyzing the key factors affecting the startup speed of CSP turbines from the perspective of bypass capacity optimization. Taking a 100MW tower CSP plant in China as the research object, the startup bypass capacity is optimized. The results show that after the optimization of the startup bypass capacity, the cold start time of the turbine is reduced by 7 minutes, which improves the startup speed, operational efficiency, equipment safety, and economic benefits of the power plant to a certain extent.

1. Introduction

Concentrated Solar Power (CSP) plants, equipped with thermal energy storage systems [1], can provide a stable power supply in renewable energy grids, compensating for the intermittent nature of photovoltaic and wind power. Their peak-shaving and frequency-regulation capabilities help mitigate grid load fluctuations, reduce wind and solar curtailment, and enhance the utilization rate of renewable energy [2]. Frequent start-stop operations and rapid response to grid dispatching are core design requirements for peak-shaving CSP plants. However, the first batch of domestic CSP demonstration projects has revealed common issues such as slow startup speeds, prolonged equipment warm-up times, significant energy losses, and imperfect control systems. These challenges make it difficult for the equipment to achieve a rapid response to grid dispatching in actual operation.

To address these issues, Jiao et al. [3] used numerical methods to identify that the thermomechanical stress of the turbine rotor is a critical factor limiting the rapid startup, frequent start-stop, and load variation of CSP turbines. Zhang et al. [4] conducted an energy analysis of the startup process of CSP plants and found that energy losses during the startup phase account for approximately half of the total input energy. They highlighted the energy-saving potential of bypass losses, which can be reduced by optimizing the startup curve.

However, previous studies have primarily focused on individual equipment and control systems. In practice, the design of the startup bypass plays a crucial role in the startup process of CSP plants [5]. It addresses the mismatch between steam generation in the evaporative heat exchange system and steam intake in the turbine, thereby reducing the time required for the turbine to transition from a cold state to full-load operation and improving system response speed. Therefore, this paper analyzes the bypass system of CSP plants from a system integration perspective, establishes a mathematical model for calculating turbine bypass capacity, and optimizes the startup bypass capacity to shorten the plant startup time. The study aims to provide a reference for the design of bypass systems in CSP turbines.

2. Analysis of the turbine bypass system in CSP plants

2.1. System composition and working principle

In a tower-type concentrated solar power (CSP) plant, solar energy is converted into thermal energy of the binary molten salt, which serves as the heat transfer fluid in the thermal energy storage system, through the receiver of the solar collector system. The thermal energy is stored in the thermal storage device and then transferred to the working fluid (steam) of the turbine power generation system via heat exchange equipment in the thermal energy storage system. Within the turbine power generation system, the steam performs work, converting thermal energy into electrical energy [6].

The turbine bypass system is a critical component of CSP plants. It adopts a two-stage series configuration consisting of high-pressure and low-pressure bypasses. The main equipment includes high-pressure bypass valves, low-pressure bypass valves, desuperheating and pressure-reducing devices, and bypass pipelines. These components work collaboratively to achieve several functions: accelerating startup speed, regulating steam flow during shutdown or load variations, protecting equipment such as reheaters during low-load operation, maintaining system pressure stability, and enhancing operational safety. A schematic diagram of the turbine bypass system in a CSP plant is shown in Figure 1[7].

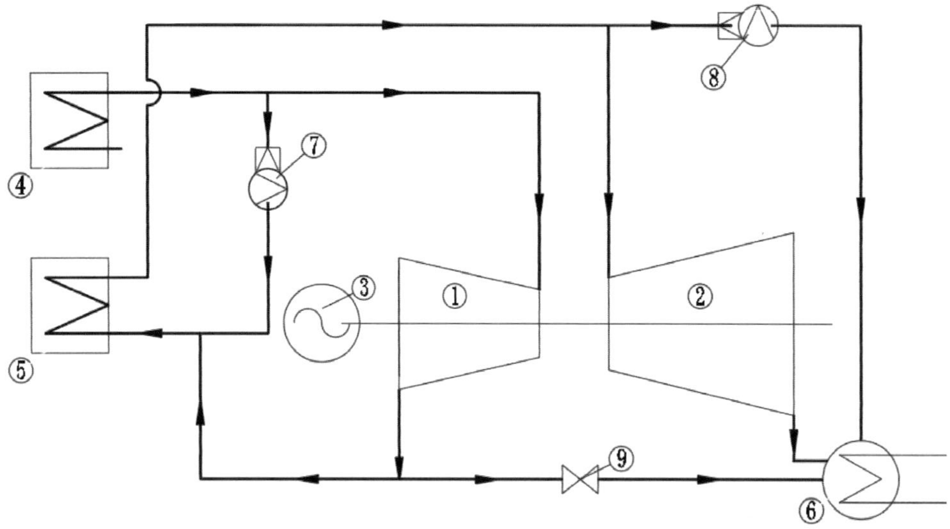

Figure 1. Schematic Diagram of the Turbine Bypass System in a CSP.
Notes: 1—High-pressure cylinder, 2—Intermediate-low pressure combined cylinder, 3—Generator, 4—Superheater, 5—Reheater, 6—Condenser, 7—High-pressure bypass valve, 8—Low-pressure bypass valve, 9—High-pressure exhaust vent valve

During startup, the high-pressure bypass system directs the main steam to bypass the high-pressure cylinder. After desuperheating and pressure reduction, the steam is channeled into the reheater [8]. Simultaneously, the low-pressure bypass system directs the reheated steam to bypass the intermediate-low-pressure combined cylinder and leads it into the condenser. After normal operation is achieved, the bypass system is gradually closed and remains in a hot standby state. During load variations or emergencies, the bypass system rapidly adjusts the steam flow to maintain system pressure stability, prevent equipment from overheating or overpressurization, and ensure system safety.

2.2. Analysis of the startup process

The variability of solar resources in concentrated solar power (CSP) plants necessitates that turbines possess rapid startup and load regulation capabilities. Since the inception of the first batch of domestic demonstration projects, CSP turbines have adopted a combined high-pressure and low-pressure cylinder startup method. Based on this startup approach, the turbine startup phase is categorized into cold startup, warm startup, hot startup, and very hot startup, depending on the temperature state of the turbine's metal components and the duration of shutdown.

Journal of Physics: Conference Series **3031** (2025) 012009 doi:10.1088/1742-6596/3031/1/012009

During a cold startup, the rotor temperature is low, resulting in the longest startup time and a slow temperature and pressure rise process. This poses a high risk of thermal stress on the turbine, indicating a mismatch between the steam generation system's output and the turbine's steam intake parameters. Consequently, the bypass system must be engaged for an extended period to strictly control the turbine's temperature rise rate and prevent overheating. From warm startup to very hot startup, the startup time gradually shortens, the risk of thermal stress decreases, and the impact on the grid is progressively reduced. A comparison of the characteristics of different startup methods for CSP turbines is presented in Table 1.

Table 1. Comparison of Characteristics of Different Startup Methods for CSP Turbines.

Startup Method	Cold Startup	Warm Startup	Hot Startup	Very Hot Startup
Shutdown Duration	≥72h	10h～72h	≤10h	≤1h
Rotor Temperature Range	<150℃	150℃～350℃	350℃～450℃	>450℃
Time to Full Load	90min	45min	20min	15min
Turbine Thermal Stress Level	High	Medium	Low	Very Low
Temperature Rise Rate Requirement	Strict control, slow temperature rise	Moderate control, faster temperature rise	Relatively fast temperature rise	Fast temperature rise
Steam Demand	Large steam volume, gradual temperature and pressure rise	Moderate steam volume, faster parameter recovery	Low steam volume, rapid parameter recovery	Minimal steam volume, rapid parameter recovery
Bypass System	Mainly operational, bypassing the turbine to accelerate heating	Partially operational, coordinating turbine heating	Minimally operational, assisting with turbine heating	Rarely operational, direct startup
Impact on the Grid	Long startup time, significant grid impact	Moderate startup time, smaller grid impact	Short startup time, minimal grid impact	Very short startup time, negligible grid impact
Equipment Protection Requirements	Strict monitoring of thermal stress to prevent equipment damage	Attention to thermal stress to avoid localized overheating	Low thermal stress risk, lower equipment protection requirements	Very low thermal stress risk, minimal equipment protection requirements

As can be seen from Table 1, the temperature level of the turbine rotor blades and the preheating efficiency directly affect the startup speed. At the same time, the capacity of the bypass system imposes constraints on the startup speed. Insufficient bypass capacity can limit the steam intake flow of the turbine, impact the load ramp rate, slow down the startup speed, and prolong the startup time. Therefore, optimizing the bypass system capacity and control strategies to accelerate steam flow regulation can, to some extent, improve the startup speed.

3. Optimization of startup capability for the turbine bypass system in CSP plants

3.1. Optimization objectives and principles

The goal of optimizing the turbine bypass system in CSP plants is to reduce the time required to transition from a cold state to full-load operation, enhance the plant's responsiveness to fluctuations in solar resources, mitigate the risk of thermal stress on equipment during startup, prevent equipment damage, minimize energy losses during startup, reduce fuel-assisted heating costs, and meet the demands of frequent start-stop operations and load variations in CSP plants.

The bypass system connects the turbine and the steam generation system, and its optimization directly impacts system complexity, safety, and operational strategies. Through a detailed analysis of the bypass operation process and rational design of the bypass system capacity, it is possible to shorten the bypass operation time and reduce energy losses without increasing system complexity or the difficulty of operational strategies. This approach achieves the optimization of the turbine's startup bypass capability.

3.2. *Analysis of bypass system operation process*

During the cold startup phase of the unit, the initial heating of the equipment is provided by the electric heater in the evaporative heat exchange system. The drum is filled with cold water, and a water circulation within the evaporative heat exchange system is established using the startup preheating pump. Once the molten salt circulation temperature reaches 280°C, the molten salt side of the evaporative heat exchange system becomes operational, facilitating heat exchange with the steam-water side to further elevate the equipment temperature and pressure. At this point, the bypass system is activated. After the evaporative heat exchange system can stably generate steam, the turbine gland sealing system and turbine pre-warming are engaged, making the turbine ready for rotation.

During the warm, hot, or very hot startup phases, the time required for the molten salt side to become operational is reduced. The subsequent startup process before turbine rotation remains consistent, although the valve opening degrees and durations vary across stages. The bypass system operation status is divided into three phases: bypass operation before turbine rotation, operation during load increase, and stable operation until bypass deactivation. Before turbine rotation, the superheater, reheater, and high/low-pressure bypass systems are simultaneously activated. The generated steam bypasses the high-pressure cylinder through the high-pressure bypass valve and reheater, and bypasses the intermediate-low pressure cylinder through the low-pressure bypass before entering the condenser. This phase continues until the main steam reaches the temperature and pressure required for turbine rotation and the superheater achieves the minimum stable steam generation rate. This phase incurs the highest cold source loss for the turbine and is the primary focus of this bypass capacity optimization. During the turbine rotation phase, the process follows the turbine startup curve. This phase involves complex bypass system operation, where part of the steam generated by the superheater passes through the high-pressure cylinder and high-pressure exhaust vent valve before entering the condenser, while the remaining steam enters the reheater via the high-pressure bypass [9]. Similarly, part of the steam generated by the reheater passes through the intermediate-low-pressure cylinder into the condenser, while the remaining steam enters the condenser via the low-pressure bypass. Subsequently, the turbine begins to increase its load until the high-pressure exhaust vent valve closes. At this point, part of the steam generated by the superheater merges with the steam from the high-pressure bypass valve after passing through the high-pressure cylinder and enters the reheater. This moment marks the peak flow rates for both the high-pressure and low-pressure bypass systems. Following this, the bypass system gradually deactivates as the turbine continues to increase its load.

3.3. *Establishment of a mathematical model for bypass capacity optimization*

High-temperature and high-pressure steam from the superheater enters the turbine and the high/low-pressure bypass systems through steam pipelines before being discharged into the condenser. To simplify the model, the following assumptions are made:

- The steam flow within the steam pipelines is considered one-dimensional.
- Heat exchange between the steam pipelines and the environment is neglected.
- Energy losses due to friction in the high-pressure and low-pressure cylinders are ignored.

The calculation of the bypass valve capacity should meet the parameters and flow requirements for cold, warm, hot, and very hot startup conditions. The operating point with the highest demand for bypass valve capacity is selected, and its parameters are then converted to the Boiler Maximum Continuous Rating (BMCR) condition. Under variable operating conditions, the conversion of bypass valve flow is calculated using Equation (1), as follows [10]:

Journal of Physics: Conference Series **3031** (2025) 012009 doi:10.1088/1742-6596/3031/1/012009

$$D_c = \frac{\sqrt{\frac{p_c}{v_c}}}{\sqrt{\frac{p_e}{v_e}}} D_e \tag{1}$$

In the equation:

D_e, p_e, v_e—represent the flow rate, pressure, and specific volume of the bypass valve when fully open under rated conditions;

D_c, p_c, v_c—represent the flow rate, pressure, and specific volume of the bypass valve when fully open under variable operating conditions.

Criteria for Critical Flow in Bypass Valves:

(1) When the pressures before and after the bypass valve satisfy Equation (2), the flow within the bypass valve is subcritical

$$\frac{p_b}{p_0} > \left[\frac{2}{k+1}\right]^{\frac{k}{k-1}} \tag{2}$$

(2) When the pressures before and after the bypass valve satisfy Equation (3), the flow within the bypass valve is critical.

$$\frac{p_b}{p_0} \le \left[\frac{2}{k+1}\right]^{\frac{k}{k-1}} \tag{3}$$

Calculation Formula for Bypass Valve Flow Area:

(1) When the flow within the bypass valve is subcritical, the flow area of the bypass valve is calculated using the following formula:

$$A = \frac{q}{3.6F_L\sqrt{\left(\frac{2k}{k-1}\right)\left[\left(\frac{p_b}{p_0}\right)^{\frac{2}{k}} - \left(\frac{p_b}{p_0}\right)^{\frac{k+1}{k}}\right]}\sqrt{\frac{p_0}{v}}} \tag{4}$$

(2) When the flow within the bypass valve is critical, the flow area of the bypass valve is calculated using the following formula:

$$A = \frac{q}{3.6F_L\sqrt{k\left(\frac{2}{k+1}\right)^{\frac{k+1}{k-1}}}\sqrt{\frac{p_0}{v}}} \tag{5}$$

In the formula

p_0—Pressure before the valve, in MPa(a);

p_b—Pressure after the valve, in MPa(a);

k—Isentropic exponent, taken as 1.3 for superheated steam and 1.135 for saturated steam;

A—Flow area of the valve, mm^2

q—Flow rate of the valve, kg/h;

F_L—Valve Flow Coefficient, (0.85～0.92)

v—Specific Volume of Steam, m^3/kg

Based on the actual design data of a 100 MW tower-type concentrated solar power (CSP) plant, the operational conditions of the bypass system, and the formulas provided in this section, the valve capacity of the bypass system is designed. At the moment when the turbine rotation closes the VV valve, the design parameters of the bypass valve under different startup conditions are shown in Table 2.

Table 2. Optimized Design Parameters of the Bypass Valve.

Item	BMCR Condition	Cold Startup	Warm Startup	Hot Startup	Very Hot Startup
Inlet Steam Pressure (Mpa. a)	14.69/3.04	6.8/0.85	6.8/0.85	8.6/0.85	8.4/0.85
Inlet Steam Temperature (℃)	548/548	420/420	486/486	515/515	542/542
Inlet Steam Flow Rate (t/h)	301/267	82/90	81/91	80/89	79/91
Flow State in Valve	Critical / Subcritical	Critical / Subcritical	Critical / Subcritical	Critical / Subcritical	Critical / Subcritical
Outlet Pressure (Mpa. a)	0.9/0.6	0.9/0.6	0.9/0.6	0.9/0.6	0.9/0.6
Outlet Steam Flow Rate (t/h)	203	90/109	91/115	89/115	91/119
Converted Flow Rate under BMCR Condition	/	54.3%	57%	45.2%	46.7%

Note: In each data entry, the left value represents the parameters of the high-pressure bypass valve, and the right value represents the parameters of the low-pressure bypass valve.

Based on the optimized design parameters of the bypass valves provided in the table above, the flow capacity of both the high-pressure and low-pressure bypass valves is set to 60% of the Boiler Maximum Continuous Rating (BMCR) condition. Accordingly, the startup curve is designed.

3.4. Impact of bypass capacity optimization on startup capability

Currently, the bypass capacity of most constructed and under-construction CSP plants in China is designed considering the bypass capacity of thermal power units with a capacity below 350 MW, i.e., the bypass capacity is set at 30%/40% of the Boiler Maximum Continuous Rating (BMCR) main stream flow. This design is based on unit capacity. However, as peak-shaving power plants, CSP plants require daily start-stop operations and are significantly affected by the variability of solar resources, leading to more frequent load changes compared to conventional thermal power units. Designing a bypass system that aligns with the actual operational strategies of CSP plants can help shorten startup times and reduce cold source losses during startup.

Based on the analysis of the bypass system operation process in Section 3.2 of this paper, optimizing the turbine bypass capacity primarily reduces the time before turbine rotation, during rotation, and load increase. The cold source loss is the highest in the first phase, while the startup duration in the latter two phases is influenced by multiple factors, such as steam pressure and temperature monitoring, vibration monitoring, and bearing temperature monitoring, often resulting in discrepancies between theoretical calculations and actual operations. Therefore, based on the actual design of a domestic 100MW tower-type CSP unit, this study ignores the long initial startup and warm-up time of the equipment, as well as external factors affecting the turbine load ramp-up after rotation. Taking cold startup as an example, the analysis starts from the introduction of high-temperature molten salt into the evaporative heat exchange system and compares the impact of bypass system capacities of 40% BMCR and 60% BMCR on the cold startup time of the turbine. Based on the above assumptions, a mathematical model of the turbine bypass system for CSP plants is established, and the cold startup curves for the 100MW CSP plant are plotted. Figure 2 shows the cold startup curve of the turbine with a bypass system capacity of 40% BMCR, and Figure 3 shows the cold startup curve with a bypass system capacity of 60% BMCR.

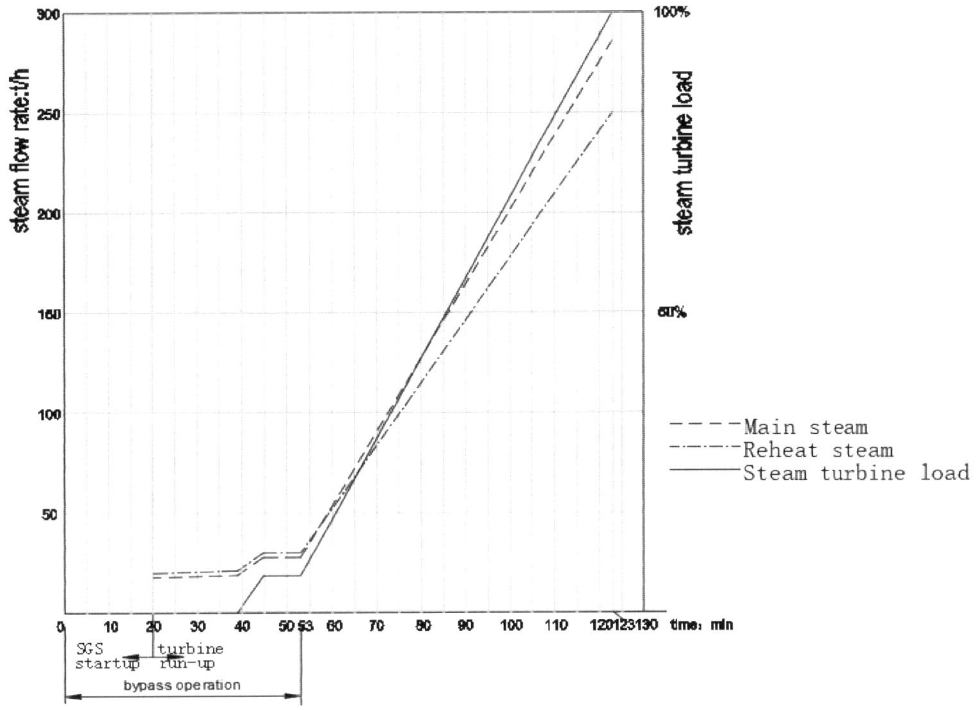

Figure 2. Cold Startup Curve of the Turbine with a Bypass System Capacity of 40% BMCR.

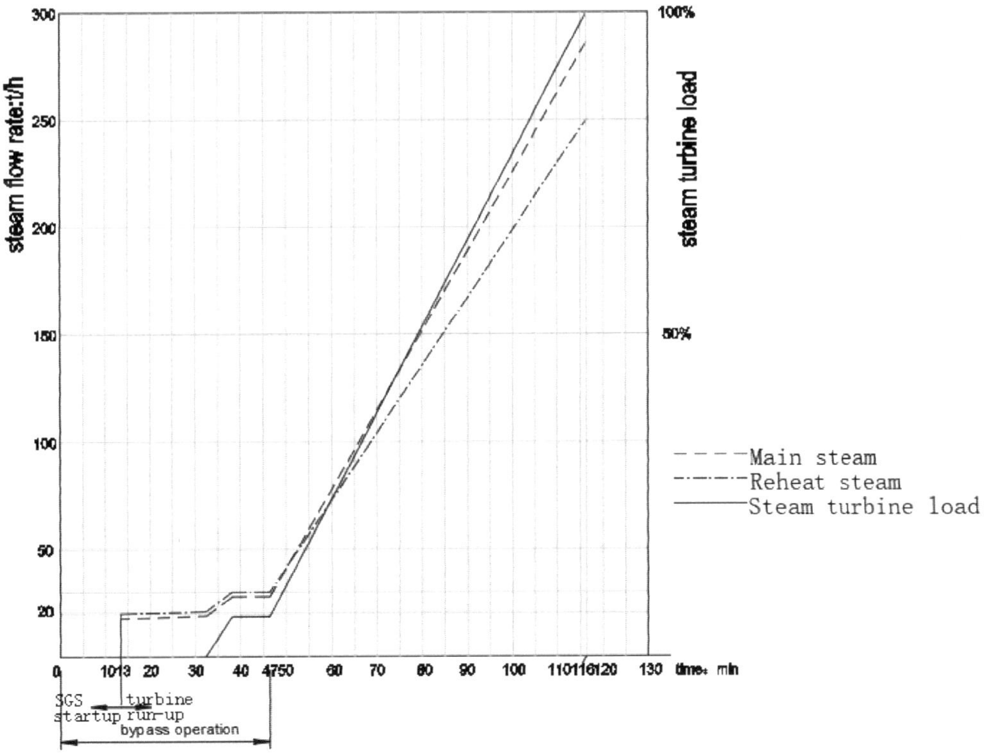

Figure 3. Cold Startup Curve of the Turbine with a Bypass System Capacity of 60% BMCR.

As can be seen from Figures 2 and 3, the startup of the CSP plant unit begins with the introduction of high-temperature molten salt into the evaporative heat exchange system. Before turbine rotation, the superheater reaches the minimum stable steam production level, and the steam production of the superheater and reheater is matched with the steam intake of the turbine. When the bypass system capacity is 60% BMCR, the cold startup takes only 13 minutes, which is 7 minutes shorter than the 40% BMCR bypass capacity, representing a 35% optimization in startup time. Currently, domestic evaporative heat exchange equipment exhibits a heating rate of 8°C/min in the temperature range of 260°C to 280°C and 10°C/min in the range of 300°C to 560°C. This means that it takes at least 9 to 10.5 minutes for the superheater to generate steam at the required temperature for turbine rotation from a cold start.

With a 40% BMCR bypass capacity, the turbine reaches full load at 123 minutes, while with a 60% BMCR bypass capacity, it reaches full load at 116 minutes. This is because once the turbine begins rotation, the startup time is primarily influenced by the combined effects of the startup strategy, turbine-generator equipment, and the bypass system, with the turbine-generator equipment being the dominant factor. Key influences include turbine thermal stress, turbine control systems, and steam parameters. Therefore, a subsequent comparison of startup capability optimization after turbine rotation under the two bypass capacities will be conducted based on actual operational data from the CSP plant in this project.

4. Conclusion

This paper addresses the optimization of the startup capability of the turbine bypass system in a 100 MW tower-type concentrated solar power (CSP) plant. Through theoretical analysis, numerical simulation, and comparison of turbine startup curve data, it is proposed that optimizing the bypass capacity can enhance the startup capability of CSP turbines, and its effectiveness is validated. Firstly, the composition of the bypass system in CSP plants and the dynamic analysis of the startup process are examined. Secondly, through a detailed analysis of the actual operation process of the bypass system, the capacity design of the bypass system is optimized to better adapt to the thermal system. Additionally, a mathematical model for capacity calculation is established by integrating the dynamic regulation capability of the bypass system, and turbine startup curves under different bypass capacities are plotted. By optimizing the bypass system capacity, the startup time before turbine rotation is significantly reduced. A bypass system with 60% BMCR capacity shortens the cold startup time before turbine rotation by 7 minutes, rapidly achieving steam balance between turbine steam intake and superheater steam production. This also reduces heat consumption during startup, minimizes turbine cold source losses, and enables the plant to achieve rapid startup.

In summary, this study optimizes the startup capability of the turbine bypass system in a 100MW tower-type CSP plant, not only improving the startup efficiency and operational stability of the plant but also reducing equipment wear and operational costs. It provides strong support for the efficient and safe operation of CSP plants. In the future, further research will be conducted on the impact of bypass capacity optimization on the actual startup process of turbines to further enhance the overall performance of CSP plants.

References

[1] Corgnale, C (2020). Techno-Economic Assessment of Destabilized Li Hydride Systems for High Temperature Thermal Energy Storage[J]. Inorganics,8(5):30.

[2] Yanming Qi, Shangang Ma, and Fubao Jin (2024). Research on Bi-level Optimal Configuration of Thermal Storage Capacity in Solar Thermal-Wind Hybrid Power Generation Systems. Thermal Power Generation,53(10):41-49.

[3] Yuxue Jiao, Zili Xu, Gang Yin and et al. (2020) Analysis of Thermomechanical Stress and Startup Optimization of Solar Thermal Power Turbine Rotors. [J]. Turbine Technology,62(06):417-420+429.

[4] Shunqi Zhang, Ming Liu, Jiping Liu, and Junjie Yan (2023). Thermodynamic Analysis of the Startup Process in Parabolic Trough CSP Plants. [J]. Journal of Engineering Thermophysics,44(03):608-616.

[5] Bo Yuan and Junfeng Lu (2022). Control Strategy and Application of Bypass System in 100 MW Parabolic Trough

Solar Thermal Power Plant. [J]. Northeast Electric Power Technology,43(05):10-13+21.
[6] Chunguang He, Wei Guo, Shunqi Zhang and et al. (2025) Research on Daily Operation Mode Optimization of Parabolic Trough Solar Thermal Power Plants Considering Startup Energy Consumption. [J]. Thermal Power Generation,54(02):88-96.
[7] Weimeng Wang, Xiaodi He, Wei Dong and Yajun Song (2024) Research on Operation Safety Protection Strategies of Steam Turbine Bypass Heating Systems. [J]. Journal of Engineering for Thermal Energy and Power,39(12):49-59.
[8] Zhang D, Zhao Y, Wu Z, et al (2019) Quantitative Analysis of Operating Economic Performance by Valve Leakage in Bypass System[J]. IOP Conference Series: Earth and Environmental Science,295(3):032091-032091.
[9] Falah Alobaid (2018) Numerical Simulation for Next Generation Thermal Power Plants[M]. Springer, Cham.pp.11-13.
[10] W S, Y W (2019) Selection of steam turbine bypass system[J]. IOP Conference Series: Earth and Environmental Science, 354012066-012066.

Journal of Physics: Conference Series **3031** (2025) 012010 doi:10.1088/1742-6596/3031/1/012010

Study on superposition sound field of filter devices used in power converter stations

Jinyu Li*, Gang Cao, Shujiao Chen, Xiaoyan Lei, Zhongqiu Zuo

China Electric Power Research Institute, Beijing 100192, China

* Corresponding author's email: ljyxjtu@sina.com

Abstract. High-Voltage Direct-Current (HVDC) transmission systems provide a critical means of delivering clean energy to economically developed areas. However, noise pollution from converter stations, particularly from filter devices such as capacitors and reactors, has become a significant concern. This paper investigates the acoustic noise generated by these filter devices, focusing on the superposition of sound fields produced by capacitors and reactors. The theoretical analysis of sound pressure superposition reveals the influence of phase differences between the sound waves emitted by each device, and experimental measurements validate these findings. By analysing sound field distributions under various excitation conditions, this paper demonstrates that interference effects between reactor and capacitor noise are prominent, significantly affecting the total noise level. The results highlight the importance of understanding the interactions between these sound sources for effective noise reduction in HVDC converter stations. This paper provides valuable insights for optimizing acoustic management strategies in power converter stations, contributing to both noise mitigation and improved environmental acoustics.

1. Introduction

High-Voltage Direct-Current (HVDC) transmission technology plays a crucial role in delivering clean energy to economically developed areas. However, the noise pollution caused by converter stations has become an increasing concern. Studies indicate that acoustic noise in power converter stations originates from filter devices, which include oil-immersed full-film capacitors and dry-type air-core reactors [1]. When a converter station operates, harmonic currents flow through these filter devices, generating electromagnetic forces that lead to vibrations and noise emissions at various frequencies [2]. Thus, investigating the acoustic noise characteristics of filter devices in power converter stations holds significant academic and engineering value.

According to the CIGRE report, primary noise sources in converter stations include converter transformers, converter valves, smoothing reactors, filter capacitor devices, and filter reactors, with filter devices reaching sound power levels as high as 105 dB [3]. Most studies on filter device noise have focused on noise generation mechanisms, estimation methods, and reduction measures [4,5]. However, there has been limited research on the superposition sound fields between capacitors and reactors. Since reactors and capacitors emit sound waves of identical frequency spectra, interference effects occur due to peak-to-peak superposition. To assess acoustic noise from filter devices, researchers have improved the signal-to-noise ratio by optimizing equipment distance and applying simple filtering techniques to obtain sound power and spectrum data [6]. However, noise from capacitors and reactors in converter stations exhibits overlapping frequency peaks, and direct noise measurements are often contaminated by other sources. This challenge necessitates further study on the superposition of sound fields from filter devices.

This paper investigates the superposition sound field of filter devices in HVDC converter stations through theoretical analysis and experimental validation. By analyzing amplitude variations in the emitted sound waves, interference occurring in practical operating conditions is studied. The findings contribute to the broader understanding of noise propagation in power converter stations and offer insights into optimizing filter device placement and operational strategies to minimize acoustic disturbances.

2. Theoretical analysis of superposition sound fields of filter devices

For two independent sound waves in a sound field, the individual sound pressures at a given receiving point are denoted as p_1 and p_2, resulting in a superposed sound pressure:

$$p = p_1 + p_2 \tag{1}$$

If the two sound waves share the same frequency and maintain a fixed phase difference, interference occurs. The sound waves arriving at the receiving point can be expressed as:

$$p_1 = p_{1a}\cos(\omega t - \varphi_1) \tag{2}$$

$$p_2 = p_{2a}\cos(\omega t - \varphi_2) \tag{3}$$

where φ is the constant phase difference between the two waves:

$$\varphi = \varphi_1 - \varphi_2 \tag{4}$$

Applying the superposition principle, the resulting sound pressure amplitude can be determined as:

$$p = p_1 + p_2 = p_{1a}\cos(\omega t - \varphi_1) + p_{2a}\cos(\omega t - \varphi_2) = p_a\cos(\omega t - \varphi) \tag{5}$$

where:

$$p_a^2 = p_{1a}^2 + p_{2a}^2 + 2p_{1a}p_{2a}\cos(\varphi_2 - \varphi_1) \tag{6}$$

$$\varphi = \tan^{-1}\frac{p_{1a}\sin\varphi_1 + p_{2a}\sin\varphi_2}{p_{1a}\cos\varphi_1 + p_{2a}\cos\varphi_2} \tag{7}$$

This equation reveals that the superposed sound pressure retains the same frequency but exhibits an amplitude dependent on the phase difference φ. In specific spatial locations around filter devices, sound waves may arrive either in phase (constructive interference) or out of phase (destructive interference), leading to significant variations in sound pressure levels.

In the space around the filter devices, if $\varphi = 0, \pm 2\pi, \pm 4\pi, \ldots$ at some positions, the two sound waves always arrive at the same phase, then

$$p_a = p_{1a} + p_{2a} \tag{8}$$

If at some position, $\varphi = \pm\pi, \pm 3\pi, \ldots$, which means that the two sound waves always arrive in opposite phases, then we have

$$p_a = p_{1a} - p_{2a} \tag{9}$$

For two sound waves with different frequencies, the combined sound pressure amplitude follows the energy relation:

$$p_a^2 = p_{1a}^2 + p_{2a}^2 \tag{10}$$

Based on the definition of sound pressure level (SPL):

$$L_{pi} = 20\lg\frac{p_i}{p_{\text{ref}}} \tag{11}$$

$$p_i^2 = p_{\text{ref}}^2 10^{L_{pi}/10} \tag{12}$$

Substituting Equation (10), the resulting SPL superposition formula is:

$$L_p = 10\lg(10^{0.1L_1} + 10^{0.1L_2}) = L_1 + \Delta L \tag{13}$$

where:

$$\Delta L = 10\lg(1 + 10^{-0.1(L_1 - L_2)}) \tag{14}$$

3. Experimental study on superposition sound fields of reactors and capacitors

3.1 Sound pressure level measurement

To evaluate the superposition sound field of capacitors and reactors, an experimental setup was established in a semi-anechoic chamber. The semi-anechoic chamber has a lower cut-off frequency of 100 Hz, and the ambient noise level of 15.5 dB at the octave band of 500 Hz. Microphones were arranged in two layers around the filter devices, as shown in Figure 1. The testing instrument and acoustic environment comply with the code given by of ISO 11201 standard [7]. According to ISO 11200 Table 1[8], the measured SPL would satisfy at least grade 2 precision, which is a 1.5 dB reproducibility standard deviation.

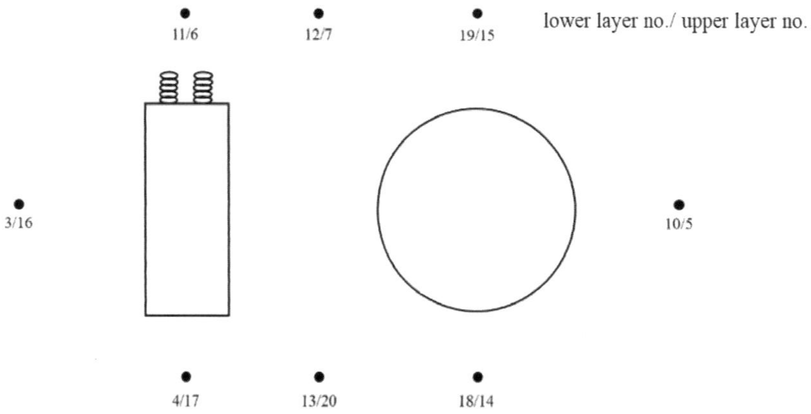

(a) Schematic diagram of the experimental platform.

(b) Picture of the interference effect experimental platform.

Figure 1. Experimental setup for measuring the superposition sound field of filter devices.

Table 1. Overview of international standards for the determination of emission sound pressure levels at the work station and at other specified positions (partial)[8]

Parameter	ISO 11201 measurement method grade 1 or 2
Test environment	Outdoors or Indoors
Criterion for suitability of test environment	indoor: – hemi-anechoic room (grade 1) – $K_2 \leq 2$ dB, (grade 2) Outdoor: – a plane area outdoorswith hard ground and noreflecting objects in thenear vicinity (grade 1) – on a reflecting planeRround ata specifieddistance from thenearest reflecting object (grade 2)
Limitation of local environmental correction K_{3A} and grade of accuracy	No determination of K_{3A} Environmentalcorrectionnot allowed

The capacitor was positioned horizontally on a support frame, with its center 0.9 m above the ground and 1.2 m from the reactor. The lower microphone was also placed 0.9 m above ground, while the upper microphone was positioned at 1.8 m to capture noise in the upper space.

According to Equations (13) and (14), when two sound sources exhibit a significant SPL difference, the superposed SPL is primarily determined by the dominant source, making interference effects difficult to analyze. Therefore, parallel excitation (applying current to the capacitor and reactor simultaneously) was used to examine interference effects.

In the experiment, a 90 A current at 550 Hz was applied to a parallel circuit consisting of the reactor and capacitor, and noise levels were measured. The reactor current was 9.42 A, and the capacitor current was 99.42 A. For independent excitation, 10 A was applied to the reactor and 100 A to the capacitor, and corresponding noise levels were recorded. The processed data led to the sound field distribution comparisons shown in Figures 2 and 3.

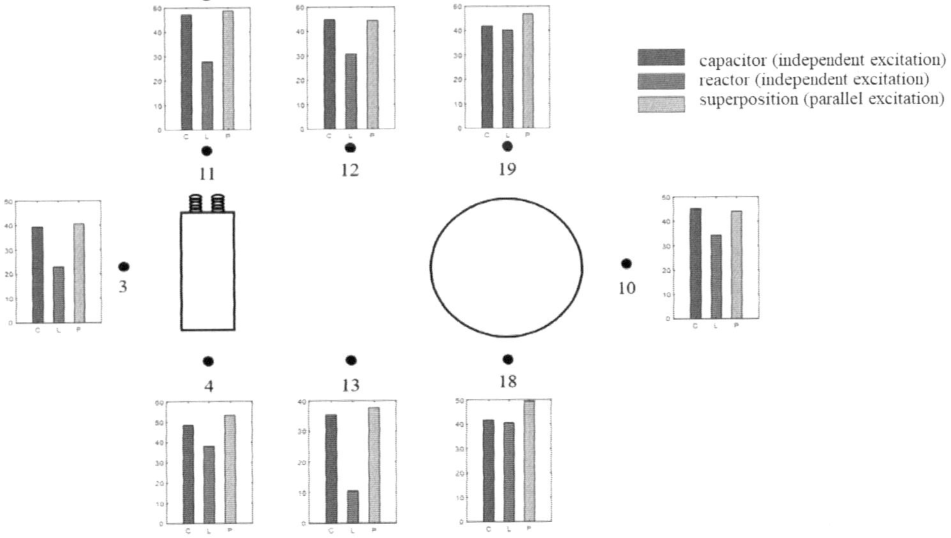

(a) Sound pressure level at the lower measurement points

Journal of Physics: Conference Series **3031** (2025) 012010
doi:10.1088/1742-6596/3031/1/012010

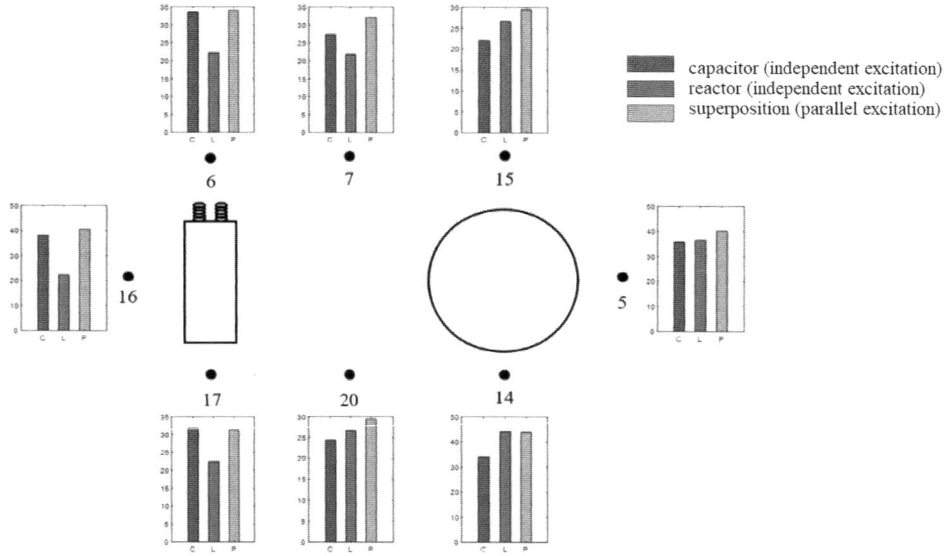

(b) Sound pressure level at the upper measurement points

Figure 2. Sound field distribution comparison at 550 Hz excitation.

To mitigate noise in HVDC converter stations, optimizing the spatial arrangement of filter devices is essential based on the interference phenomenon in the sound field. Increasing the physical distance between reactors and capacitors can minimize constructive interference effects arising from the superposition of sound waves at identical frequencies, thereby reducing overall noise levels. Additionally, strategic positioning of these components to induce phase differences that promote destructive rather than constructive interference can further suppress resultant sound pressure levels. Such spatial modifications contribute to improved acoustic management by minimizing noise amplification caused by wave superposition effects.

3.2 Analysis of the superposition sound field from filter devices

Theoretical SPL values were computed using Equation (10) and compared with experimental results to evaluate interference effects. A measurement location with a significant SPL difference was chosen for incoherent superposition calculations, as shown in Figure 3.

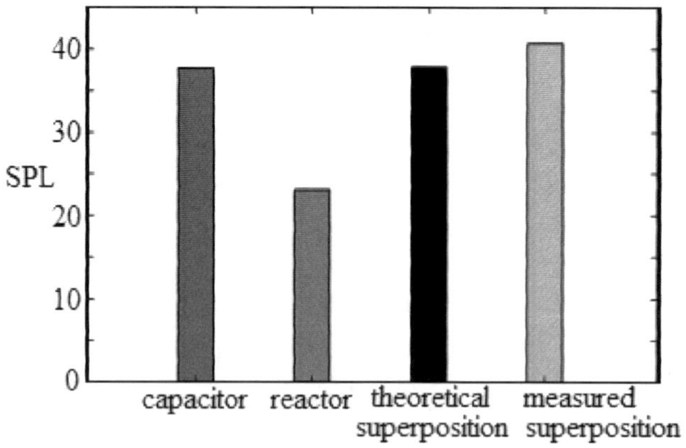

Figure 3. Comparison between theoretical calculation and actual measurement at Position No. 3 (large SPL difference).

When noise levels significantly differed, the superposed sound field was primarily influenced by the dominant source. The measured SPL exceeded the theoretical incoherent superposition value. Specifically, when the capacitor was independently excited, the measured SPL was 37.69 dB; when the reactor was independently excited, it yielded 23.09 dB. The theoretical superposition value was 37.85 dB, whereas the actual measured value was 40.68 dB. So, 2.83 dB higher than expected.

Another measurement position, where the capacitor and reactor SPLs were approximately equal, was analyzed (Figure 4).

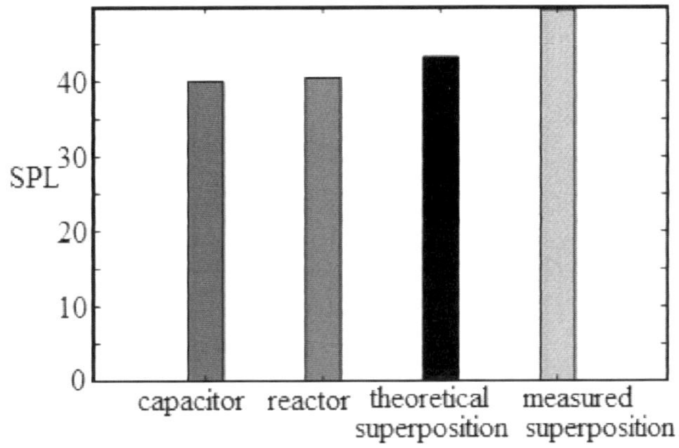

Figure 4. Comparison between theoretical calculation and actual measurement at Position No. 18 (approximately equal SPLs).

When SPLs were similar, the superposition effect was more pronounced. Independent excitation measurements recorded SPLs of 41.05 dB (capacitor) and 41.53 dB (reactor). The theoretical calculation predicted 44.81 dB, while the measured value reached 48.62 dB, so 3.81 dB higher. This confirms that interference effects significantly impact the superposition sound field, and the combined SPL is not merely an energy-based superposition of the two sources.

4. Conclusion

This paper analyzed the superposition sound field of capacitors and reactors in HVDC converter stations. Theoretical calculations and experimental measurements demonstrated that the superposed SPL is influenced by phase differences and relative amplitudes of the sound sources. The findings provide critical insights for noise reduction strategies in power converter stations, contributing to improved acoustic environmental management.

Acknowledgments

The investigation was supported by the CEPRI project: acoustic performance evaluation technology of dry-type air-core reactor based on electromagnetic-structural-acoustic coupling fields (SZ83-23-001; 524200230025).

References

[1] Zhang D, Zhang Y, Zhao Y, Huang Y, Xiao K. (2021) Research on noise reduction measures for dry-type air-core filter reactors in ±800 kV DC transmission projects. Acoustic Technology, vol.4: 5-23.

[2] Smede H, Johansson C G, Winroth O, Schutt, H. P. (1995) Design of HVDC converter stations with respect to audible noise requirements. IEEE Transactions on Power Delivery, vol.2: pp. 747–758.

[3] CIGRE. (2002) HVDC Stations Audible Noise. CIGRE technical reports. 6-32–6-39.

[4] Chen S, Tian J, Liang Z, Shu T, Gao L. (2023) Research on interference in noise measurement of dry air-core reactors. High Voltage Electrical Appliances, vol.2: pp.147-152.

[5] Hou S, Wang L, Yang F. (2023) Vibration noise analysis and control of bridge arm reactor. High Voltage, vol.06: pp.2385-2396.
[6] Jiang T, Zhe Y. (2024) Comparative Analysis of Magnetic Field Distribution Characteristics of Two Shapes of Air-Core Bridge Arm Reactors. Energies, vol.17, no.18: 4652-4655.
[7] International Organization for Standardization. (2010) Acoustics — Noise emitted by machinery and equipment — Determination of emission sound pressure levels at a work station and at other specified positions in an essentially free field over a reflecting plane with negligible environmental corrections. ISO 11201:2010.
[8] International Organization for Standardization. (2014) Acoustics — Noise emitted by machinery and equipment — Guidelines for the use of basic standards for the determination of emission sound pressure levels at a work station and at other specified positions. ISO 11200:2014.

Journal of Physics: Conference Series **3031** (2025) 012011 doi:10.1088/1742-6596/3031/1/012011

Feasibility of cooling the steam turbine with the remaining heat of the nuclear island

Ding Zhang[1*], Bohong Zhang[2], Ze Zhang[1]

[1] Suzhou Nuclear Power Research Institute, Suzhou 215000, China
[2] Guangxi Fangchenggang Nuclear Power Co., Ltd, Fangchenggang, 530000, China

[*]Corresponding author's e-mail: 1273964273@qq.com

Abstract. The maintenance of the nuclear power unit needs to cool the steam turbine. The fast cooling of the steam turbine can effectively shortcut the cooling duration. Many fast cooling systems use the steam supplied by the nuclear island to rotate the steam turbine. Therefore many nuclear plants start the shutdown process of the nuclear island only after the steam turbine has been cooled. If the remaining heat of the nuclear island can meet the demand of the fast cooling system, the shutdown process of the nuclear island and the fast cooling of the steam turbine can be done at the same time. This means using the remaining heat of the nuclear island to cool the steam turbine can reduce the repair time. To evaluate the feasibility of cooling the steam turbine with the remaining heat of the nuclear island, this paper establishes a method of calculating the remaining heat of the nuclear island through the thermodynamic model of the secondary circuit. A model for estimating the power required by the fast cooling system is also established. With the help of the model established in this paper, it preliminary verifies that the remaining heat of the nuclear island can meet the demand of the fast cooling system in a nuclear power plant. The method proposed in this paper can also be applied in other nuclear power plants and help reduce the repair time.

1. Introduction

Nuclear power plants serve as a significant source of energy, providing a substantial amount of electricity to meet global demand[1]. Additionally, they play a crucial role in reducing reliance on fossil fuels and mitigating environmental impacts[2-3]. Turbines are an important component of nuclear power plants[4-7]. In the overhaul of the nuclear power unit, the fast cooling of the steam turbine can make the temperature of various parts of the steam engine drop as soon as possible to meet the maintenance conditions[8-10]. Many nuclear power plants use the steam supplied by the nuclear island to rotate the steam turbine. Therefore it often starts the shutdown process of the nuclear island when the steam engine is cooled. If the fast cooling of the steam turbine can use the remaining heat of the nuclear island to rotate the steam turbine, it can conduct the shutdown process of the nuclear island and the fast cooling of the steam turbine at the same time. In other words, using the remaining heat of the nuclear island to cool the steam turbine can reduce the repair time.

This paper establishes a model which can be used to calculate the remaining heat of the nuclear island and the power required by the fast cooling system. With the help of the model established in this paper, it can evaluate the feasibility of cooling the steam turbine with the remaining heat of the nuclear island.

2. Model

Data within 3 hours after the turbine shutdown is used to calculate the remaining heat of the nuclear island. The model needs to find a period when the ASG flow rate is steady, then the remaining heat of the nuclear island can be obtained by the following equation:

$$P_r = H_{steam}(m_{ASG} - m_{\text{pollution}}) + H_{\text{pollution}} m_{\text{pollution}} \\ -H_{\text{feedwater}} m_{\text{feedwater}} - (H_{steam} - H_{\text{sewage}}) m_{SG} \tag{1}$$

where H_{steam}, $H_{\text{pollution}}$, and H_{water} are the enthalpy of the main steam, SG sewage and the water in the SG, m_{ASG}, $m_{\text{pollution}}$ and $m_{\text{feedwater}}$ are the flow rate of the ASG feedwater, SG sewage and the SG feedwater, m_{SG} is the mass change of the water in the SG.

The power required by the fast cooling system is a linear function of the resistance torque:

$$P_c = aM_{turbine} \tag{2}$$

where M_{turbine} is the resistance torque of the steam turbine and a is a coefficient. The parameter a is obtained by the start-up test of non-nuclear steam.

The key issue is to obtain the resistance torque of the steam turbine. In this paper, the resistance torque of the steam turbine is assumed to be a linear function of the temperature of the sea:

$$M_{turbine} = bT_{sea} \tag{3}$$

where *T*sea is the temperature of the sea and b is the coefficient. This assumption will be proved in the next section.

3. Results and discussion

3.1. Influence of the sea temperature on the resistance torque

The time consumed to reduce speed from 1500 r min-1 to 1350 r min-1(angular acceleration) is used to verify that the equation (3) is reasonable. As shown in Figure 1, Angular acceleration is positively correlated with seawater temperature with good linearity. U1 and U2 in this paper mean the No. 1 unit and No. 2 unit of the nuclear power plant which needs to verify the feasibility of cooling the steam turbine with the remaining heat of the nuclear island. U0 is the data collected from another nuclear power plant. The angular acceleration is a linear function of the resistance torque, therefore it is suitable to assume that the resistance torque of the steam turbine is a linear function of the temperature of the sea.

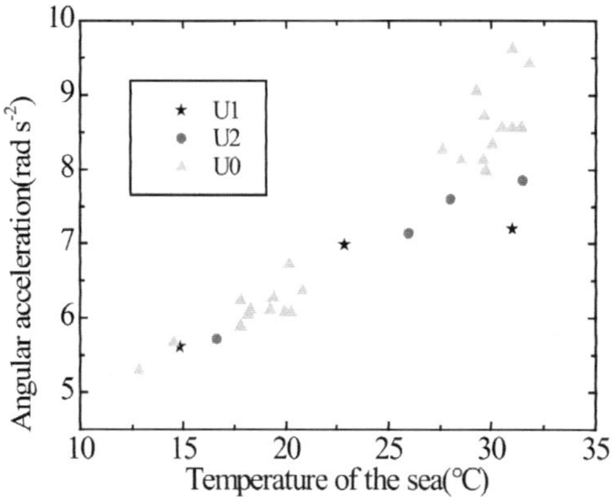

Figure 1. The relationship between the temperature of the sea and the angular acceleration.

The feasibility of Equation (3) can also be verified by the relationship between the temperature of the sea and the Increment of the valve opening. When the turbine valves are closed, it needs to increase the GCT valve opening. The reason for increasing the GCT valve opening is to make the steam passing through the turbine flow into the cooling tower through the GCT valve. Therefore, the increase of the GCT valve opening is a linear function of the resistance torque of the steam turbine. The relationship between the temperature of the sea and the increment of the GCT valve opening can be found in Figure

2. Clearly, the increment of the GCT valve opening increases with the temperature of the sea. Combining the previous analysis, the resistance torque of the steam turbine can be calculated by a linear function of the sea temperature.

As shown in equation (2), the resistance torque of the steam turbine is the product of angular velocity and moment of inertia. Therefore it still needs to obtain the moment of inertia of the steam turbine. Because the turbines in this article are all the same model and manufactured by the Alstom corporation, the moment of inertia of the steam turbine is set as a constant value. The moment of inertia of the steam turbine is obtained from the manual provided by the manufacturer, the value is 566758 kg m2.

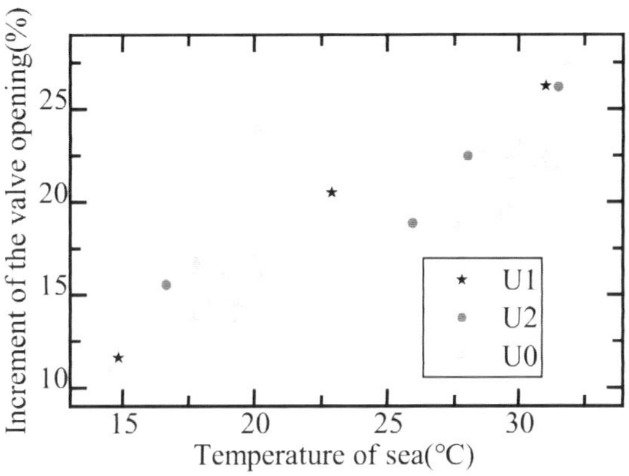

Figure 2. The relationship between the temperature of the sea and the increment of the GCT valve opening.

3.2. The correction factor of Pc

The results in the previous section can be used to calculate the resistance torque of the steam turbine. The energy required to overcome the resistance can be obtained by the resistance torque and the rotational speed, however, the power required by the fast cooling system does not equal the energy required to overcome resistance because of the heat loss. Therefore it needs to obtain a correction factor to calculate the power required by the fast cooling system.

The efficiency of the high-pressure cylinder cannot be obtained during the no-load period of the turbine, therefore the enthalpy drop of steam cannot be used to calculate the power output by the high-pressure cylinder. It can be assumed that the efficiency of the turbine is stable during this process, and the power output by the new steam is proportional to the flow rate and also proportional to the thermal power of the first circuit. In the start-up test of non-nuclear steam, a little steam flows through the GCT valve to enter the condenser and the heat power of the one circuit that can maintain the no-load fast cooling of the turbine. Therefore the data of the start-up test of non-nuclear steam can be used to obtain the correction factor of Pc.

The parameter a in Equation (2) is obtained by the start-up test of non-nuclear steam. We have obtained the data of two start-up tests of non-nuclear steam, the test results can be found in Table 1. It is worth noting that the seawater temperature is close to that of the two conditions, the back pressure values are slightly different, and the deviation of the thermal power required to maintain 1500 r/min idling is larger. It is worth noting that there is a certain energy loss when the first loop thermal power is converted to the second loop thermal power, and according to the measurement test of the first loop thermal loss, the thermal loss of the first loop during the thermal shutdown period is 3.125 MW.

With the help of Equation (2) and the data in Table 1, it can obtain the relationship between the temperature of the sea and the power required by the fast cooling system, as shown in Figure 2. The results obtained by the ref1 is higher than that obtained by the ref2. When the temperature of the sea is 34°C, the power required by the fast cooling system calculated by ref1 is 29.4MW.

Journal of Physics: Conference Series **3031** (2025) 012011 doi:10.1088/1742-6596/3031/1/012011

Table 1. Results obtained from the start-up test of non-nuclear steam.

parameter	*unit*	*Ref1*	*Ref2*
power	MW	34.01	27.28
temperature of the sea	°C	27.41	27.22
resistance torque	(N*m)	6188306	6167586

The relationship between the temperature of the sea and the power required by the fast cooling system can be found in Figure 3. The data of ref2 is much larger than the data of the ref1. This can be explained by the vacuum tightness of the condenser. The performance of the vacuum used in ref2 is poor and back pressure is high, therefore it needs more power to overcome the rotational resistance.

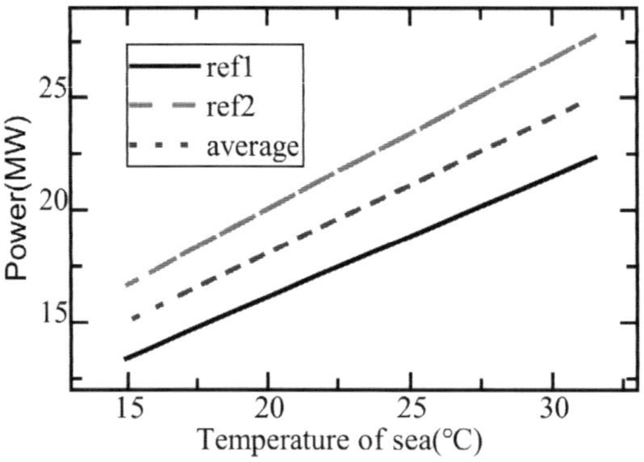

Figure 3. The relationship between the temperature of the sea and the power required by the fast cooling system.

3.3. *Feasibility of the fast cooling method*

The data within 3 hours after turbine shutdown is listed in Table 2. The flow rate of the SG sewage is larger than 0 kg/s, which means that the drain sewage is not completely isolated in these overhaul processes. The heat carried by the sewage is larger than 5MW. If the sewage is not discharged, this heat can be used to maintain the rotation of the steam turbine.

The remaining heat of the nuclear island calculated by equation (1) is also plotted in Figure 4. The calculated remaining heat is concentrated between 29.95 and 33.61 MW after 3 h of each trip. It needs to be pointed out that, drain sewage is not completely isolated in these overhaul processes. Therefore it can be considered that the remaining heat of the nuclear island is larger than 29.95MW.

The power of the electric heater in the pressure stabilizer is about 1.4 MW. This electric heater can also be used to provide energy for the fast cooling of the steam turbine. Improving the vacuum of the condenser can decrease the rotational resistance. Considering that the remaining heat of the nuclear island is larger than the power required by the fast cooling system, it can be judged that it is feasible to carry out no-load fast cooling of the turbine according to the new idea, and it can be maintained for 3h.

Table 2. The heat Data within 3 hours after turbine shutdown.

	U103	U104	U202	U203	U204	U205
H_{steam}(kJ/kg)	2767	2766	2767	2767	2767	2766
$H_{\mathrm{pollution}}$(kJ/kg)	1284	1289	1285	1286	1285	1288
H_{water}(kJ/kg)	139	176	135	160	119	165
m_{ASG}(kg/s)	18.6	15.0	13.7	14.6	18.0	15.1
$m_{\mathrm{pollution}}$(kg/s)	10.7	9.5	6.0	6.2	8.7	4.0
$m_{\mathrm{feedwater}}$(kg/s)	18.6	15.0	13.7	14.6	18.0	15.1
m_{SG}(kg/s)	0.58	3.42	4.12	2.62	1.19	-0.12
P_{r}(MW)	32.2	29.9	33.2	32.8	33.1	33.6

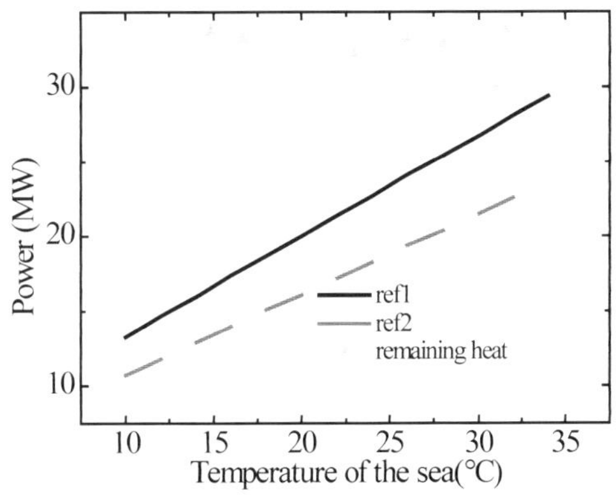

Figure 4. The power required by the fast cooling system and the remaining heat of the nuclear island.

4. Conclusion

A model is established in this paper to calculate the remaining heat of the nuclear island and the power required by the fast cooling system. The test data obtained from two nuclear power plants is used to obtain the parameters of the model. With the help of the model developed in this paper, the remaining heat of the nuclear island and the power required by the fast cooling system in the nuclear power plant are calculated. The results show that the power required by the fast cooling system calculated is smaller than 29.4MW. The remaining heat of the nuclear island is larger than 29.95. Closing the sewage discharge, opening the electric heater in the pressure stabilizer, and increasing the vacuum of the condenser can provide additional support for the fast cooling system. Therefore the remaining heat of the nuclear island can meet the demand of the fast cooling system.

References

[1] Gaurav K , Singh B K , Kumar V .Intelligent fault monitoring and reliability analysis in safety–critical systems of nuclear power plants using SIAO-CNN-ORNN[J].Multimedia Tools & Applications, 2024, 83(22).
[2] Banaszkiewicz M , Skwaro M .Numerical investigations of transient thermal loading of steam turbines for SMR plants[J].Archives of Thermodynamics, 2023, 44(4).
[3] Zhang Q .Pressure Transient Analysis on the Condenser of the HPR1000 Nuclear Power Unit[J].Energies, 2024, 17(5).
[4] Li S , Qu T , Wang D .The Impact and analysis of nuclear power plant's steam extraction and heating modification on the turbine[J].Journal of Physics: Conference Series, 2024, 2757(1).
[5] Zhou J , An Z , Yang Z ,et al.PT-Informer: A Deep Learning Framework for Nuclear Steam Turbine Fault Diagnosis and Prediction[J].Machines, 2023, 11(8).
[6] Banaszkiewicz M , Skwaro M .Numerical investigations of transient thermal loading of steam turbines for SMR plants[J].Archives of Thermodynamics, 2023, 44(4).
[7] Kong D , Liu W , Li Z ,et al.Numerical investigation on seismic response of isolated APR1400 main steam pipe under optimized support direction[J].International Journal of Pressure Vessels and Piping, 2024:209.
[8] Wang X , Zhao G , Xinhe Q U ,et al.Investigating the off-design performance of the secondary circuit system in a small modular reactor nuclear power plant[J].Journal of Tsinghua University (Science and Technology), 2024, 64(1).
[9] Xin W U , Lei L I , Zhengming W ,et al. Analysis on Rapid Cooling Technology in a 1000 MW Steam Turbine and its Application[J]. Electric Power, 2019, 52(2).
[10] JIANG Lei. Exploration of Turbine No-Load Cooling Standard in Hongyanhe Nuclear Power Unit's Overhaul[J]. Northeast Electric Power Technology, 2021, 42(10).

Journal of Physics: Conference Series **3031** (2025) 012012 doi:10.1088/1742-6596/3031/1/012012

Critical proportion of thermal power with wind speed forecast errors in fast frequency response

Junjie Huang*, **Anjia Mao**

School of Electrical and Electronic Engineering, North China Electric Power University, Changping District, Beijing 102206, China

*Corresponding author's e-mail: h759797432@163.com

Abstract. The low inertia of high renewable energy systems challenges grid frequency stability. Virtual inertia control of wind turbines improves response performance. A method is proposed using kernel density and Latin hypercube sampling to calculate wind farm inertia, considering wind speed variations and prediction errors. The Rate of change of frequency (ROCOF) models inertia demand and a critical thermal power proportion are derived for stability. MATLAB/Simulink validates the approach with a 3-machine, 9-bus model, confirming accuracy and effectiveness.

1. Introduction

To address the issue of insufficient system inertia, research on system inertia demand has emerged. Previous studies on system inertia demand have adopted diverse approaches. Ma N et al. [1] established a simulation model considering spatiotemporal characteristics and frequency response constraints at nodes to determine the minimum inertia. Soni N et al. [2] focused on the inertia response process and energy conversion, taking into account factors such as load static voltage to develop an evaluation model. Ashabani S M et al. [3] proposed a method for microgrids in different modes to enhance inertia situational awareness. However, these studies overlooked the impact of environmental fluctuations on renewable energy units' inertia response, which is addressed in this study.

Previous studies overlooked the impact of environmental fluctuations, such as solar irradiation and wind speed, on renewable energy units' inertia response, assuming rated conditions, which overestimated system inertia[4]. This study derives the minimum inertia demand expression considering environmental fluctuations and calculates the critical proportion of thermal power required to meet this constraint, aiding operators in ensuring power system safety and stability.

2. Relationship between wind turbine inertia and wind speed

Below the rated wind speed, the time-varying function $C_p(t,\beta)$ of the wind power coefficient is simplified as follows:

$$C_{\mathrm{p}}(t,0) = 0.22\left(116\lambda^{-1}(t) - 9.06\right)\mathrm{e}^{-12.5\lambda^{-1}(t)+0.44} \tag{1}$$

In the equation: C_p represents the wind power coefficient; λ denotes the tip-speed ratio; and t represents time in seconds (s).

When the wind speed exceeds the rated wind speed, the wind turbine operates in the constant rotational speed and constant power regions [5]. Due to pitch angle control, $C_p(t,\beta)$ is calculated as a function of the blade angular velocity and pitch angle. In the constant rotational speed region, the pitch angles corresponding to different wind speeds are recorded. In the constant power region, the pitch angles under high wind speeds are obtained by back-calculating while maintaining the power constant at $P_{\max}$. A function fitting the pitch angle to wind speed is derived and substituted into the mechanical power function. The subsequent steps are consistent with those in the Maximum Power Point Tracking (MPPT) region.

The virtual inertia response of a wind turbine provides short-term additional active power support to the grid [6]. This process is divided into two stages: virtual inertia response (from t_{start} to t_{end}) and rotor kinetic energy recovery (from t_{end} to $t_{recover}$), as shown in Figure 1.

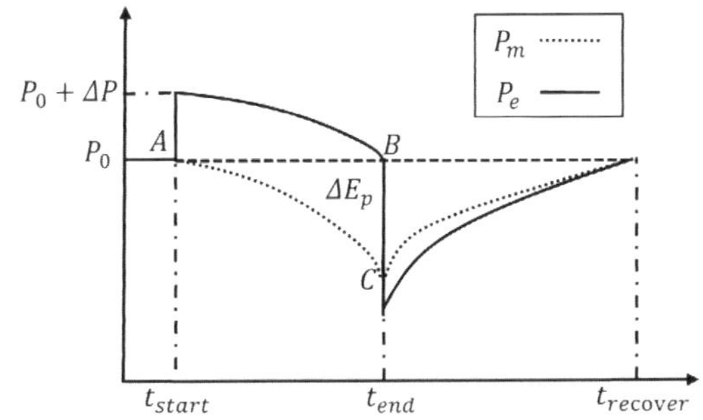

Figure 1. Wind turbine virtual inertia response.

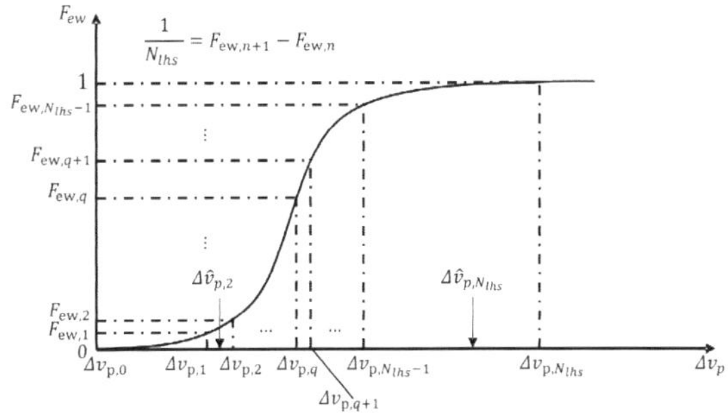

Figure 2. Latin Hypercube Sampling Principle.

When frequency fluctuations occur in the system, the wind turbine's virtual inertia control responds, causing the rotor speed to decrease from its initial speed ω_0 to the critical speed $\omega_{\min}$ for inertia response [7]. After the wind turbine completes its virtual inertia response, the system proceeds with primary and secondary frequency responses, during which the rotor speed ω_r gradually recovers to the optimal speed under the current wind speed conditions [8].

In this study, the evaluation of wind turbine inertia under fluctuating wind speed conditions depends solely on the virtual inertia response stage. The area enclosed by points A, B, and C in Figure 1 represents the kinetic energy loss of the wind turbine rotor ΔE_{p} during the inertia response stage, which provides frequency support during power disturbances. ΔE_{p} can be equivalently expressed as the virtual inertia of the wind turbine H_{wk} participating in inertia response. ΔE_{p} can be represented by the following equation:

$$\Delta E_{\mathrm{p}} = -\int_{t_{\mathrm{start}}}^{t_{\mathrm{end}}} [P_{\mathrm{m}}(t) - P_{\mathrm{m}}(0)]\mathrm{d}t \tag{2}$$

Combining Equations (1) and (2), the virtual inertia of the wind turbine participating in the inertia response is expressed by Equation (3). It can be observed that, in actual operation, wind turbine inertia changes with wind speed due to the variability of wind speed, which does not remain constant at the rated value.

$$\begin{aligned} H_{wk} &= \frac{\Delta E_{\mathrm{p}}}{S_{wk}} = \frac{-\int_{t_{\mathrm{start}}}^{t_{\mathrm{end}}} (\frac{1}{2}\rho A C_{\mathrm{p}}(t,\beta) V_W^3 - P_0)\mathrm{d}t}{S_{wk}} \\ &= \frac{-\int_{t_{\mathrm{start}}}^{t_{\mathrm{end}}} \left[\frac{1}{2}\rho A (C_{\mathrm{p}}(t,\beta) - C_{\mathrm{p}}(0,\beta)) V_W^3\right]\mathrm{d}t}{S_{wk}} \end{aligned} \tag{3}$$

In the equation: S_{wk} represents the rated capacity of wind turbine, in megawatts (MW); H_{wk} denotes the equivalent inertia of a wind turbine under actual wind speed conditions, in seconds (s).

2.1. Estimation of wind turbine inertia-considering wind speed prediction errors

This study proposes the use of the widely applied Latin hypercube sampling (LHS) method for scenario sampling. Based on the probabilistic distribution characteristics of wind speed prediction errors, a representative set of samples is extracted to replace complex historical data, as illustrated in Figure 2.

LHS is a uniform sampling method in a multidimensional space. First, the sampling process focuses on a single dimension of wind speed prediction errors, with the total number of samples denoted as N_{sl}, determining the dimensionality of the LHS as 1. The cumulative probability distribution curve is then divided into N_{sl} intervals along the vertical axis (cumulative probability) with equal spacing, where the spacing is $1/N_{lhs} = F_{ew,n+1} - F_{ew,n}$, resulting in $N_{lhs}+1$ boundary points: $0, F_{ew,1}, \ldots, F_{ew.N_{lhs}-1}, 1$. Similarly, the horizontal axis is divided into N_{sl} intervals, with boundary points denoted as $\Delta v_{p,0}, \Delta v_{p,1}, \ldots, \Delta v_{p,N_{lhs}}$ ($\Delta v_{p,0}$ represents the smallest wind speed prediction error among the N_{lhs} original data samples). A random point is selected within each interval on the horizontal axis as the sampling result for that interval. As shown in Figure 2, the sampling points are denser in regions where the slope of the cumulative probability distribution curve is steeper (indicating higher probability density), adhering to the sampling principles.

For the probability density function of wind speed prediction errors, this study applies the following approach. Historical wind speed prediction data from the wind farm are binned according to forecasted wind speed levels, and the distribution characteristics of wind speed prediction errors within each bin are statistically analyzed. Considering the versatility and accuracy of kernel density estimation (KDE) as a non-parametric estimation method, this study employs KDE to calculate the probability density function of wind speed prediction errors within each bin. The principle of KDE involves overlaying a kernel function at each statistical sample point to approximate the overall probability distribution. The normal distribution probability density function is adopted in this study as the kernel function. The KDE process is as follows:

$$\begin{aligned} & g_{\mathrm{ew},q}(\Delta v_p, \Delta\hat{v}_{\mathrm{p},q}) \\ & = \frac{1}{N_{lhs}}\frac{1}{\sqrt{2\pi}\sigma}\cdot\exp\left(-\frac{(\Delta v_p - \Delta\hat{v}_{\mathrm{p},q})^2}{2\sigma^2}\right) \\ & q = 1, 2, \cdots, N_{lhs} \end{aligned} \tag{4}$$

$$f_{ew}\left(\Delta v_{\mathrm{p}}\right) = \sum_{q=1}^{N_{lhs}} g_{\mathrm{ew},q}\left(\Delta v_p, \Delta\hat{v}_{\mathrm{p},q}\right) \tag{5}$$

In the equation: N_{lhs} represents the number of samples; Δv_{p} denotes the actual wind speed prediction error; $\Delta\hat{v}_{\mathrm{p},q}$ refers to the wind speed prediction error corresponding to the q-th sample; σ represents the parameter of the normal distribution; $g_{\mathrm{ew},q}$ refers to the kernel function for the q-th sample; $f_{ew}\left(\Delta v_{\mathrm{p}}\right)$ denotes the probability density function of wind speed prediction errors obtained from kernel density estimation. Furthermore, the cumulative probability distribution function corresponding to the wind speed prediction errors, denoted as $F_{ew}\left(\Delta v_{\mathrm{p}}\right)$, is calculated. By combining the sampled data, the probability assessment of wind speed prediction errors can be completed.

If the forecasted wind speed is V_W, based on the cumulative distribution function of wind speed prediction errors, $F_{ew}\left(\Delta v_{\mathrm{p}}\right)$, N_{lhs} random samples are extracted. For the Latin hypercube sampling data, the confidence interval corresponding to a confidence level of $\theta_k, k = 1,2,3$ is selected. Accordingly, the minimum forecasted wind speed considering prediction errors, $\mathrm{V}_{W\mathrm{mi}}$, and the maximum forecasted wind speed, $\mathrm{V}_{W\mathrm{ma}}$, are obtained for different confidence levels.

$$[\mathrm{V}_{W\mathrm{mi}}, \mathrm{V}_{W\mathrm{ma}}] = [F_{ew}^{-1}(\frac{1-\theta_k}{2}), F_{ew}^{-1}(\frac{1+\theta_k}{2})] \tag{6}$$

The minimum forecasted wind speed, $\mathrm{V}_{W\mathrm{mi}}$, is substituted into Equation (3) to obtain the minimum inertia, $H_{t,w}$, for different confidence levels.

2.2. Derivation and expression of the critical thermal power proportion under RoCoF constraints

In traditional power systems, the inertia typically doesn't require additional consideration. However, with the decreasing generation proportion of conventional regulating units such as thermal power, the system's equivalent inertia time constant may fall below the actual system requirements [9]. This issue may persist even when renewable energy units provide inertia support [10]. Therefore, this section presents a theoretical derivation of the relationship between the system's inertia demand and the generation proportion of thermal power units. Wind turbine inertia modeling is under different wind speed conditions.

Starting from the definition of the inertia time constant of generating units, the following equation can be derived [11].

$$H_{tsys} = \frac{P_t}{2RoCoF\sum S_{t,N}} f_{sys} \tag{7}$$

In the equation: H_{tsys} represents the system's minimum inertia demand in seconds (s); $\sum S_{t,N}$ denotes the rated capacity of the generating units in operation at time t, in megawatts (MW); and f_{sys} represents the system frequency in hertz (Hz).

In a system considering wind turbine inertia contribution, the equivalent inertia time constant of the system at any time t is expressed as

$$H_{real} = \frac{\sum_i S_i \cdot H_{\mathrm{G}i} + \sum_k S_k \cdot H_{t,Wk}}{\sum_i S_i + \sum_k S_k} \tag{8}$$

In the equation: i represents the number of thermal power units; k represents the number of wind power units; $H_{\mathrm{G}i}$ denotes the inertia of the *i*-th thermal power unit in seconds (s); $H_{t,Wk}$ denotes the inertia of the k-th wind power unit during inertia response at time t in seconds (s).

Assuming the inertia of each thermal power unit is identical, denoted as H_{G}, and the inertia of each wind power unit is identical, denoted as $H_{t,W}$, the system's reserved capacity, ξ, is expressed as follows:

$1-\xi = \frac{S_{t,N} - S_R}{S_{t,N}} = 20\%$,assuming $H_{tsys} = H_{real}$,

$$k_{f\lim} = \frac{P_t f_{sys} R_e - 2R_{\mathrm{m}} H_{t,W}\xi}{2R_{\mathrm{m}}(\xi H_{\mathrm{G}} - H_{t,W})\xi - (1-R_e)P_t f_{sys}} \tag{9}$$

In the equation: $k_{f\lim}$ represents the critical generation proportion of thermal power units; R_m denotes the system's maximum frequency variation rate constraint.

It should be noted that when $H_{t,W} \geq (P_t f_{sys} R_e)/(2RoCoF_{\max}\xi S_R)$, the result of the expression for $k_{f\lim}$ is less than or equal to 0 or greater than 1, which indicates that the system can be entirely composed of wind power units with supplemental inertia control.

The variation of the difference u between H_{tsys} and H_{real} is evaluated with respect to the thermal power proportion and power imbalance. The intersection line between the plane u=0 and the surface formed by the difference u of H_{tsys} and H_{real} represents the curve of the critical thermal power generation proportion as a function of power imbalance.

To validate the rationality of the conclusion, the wind speed is set to 8 ms^{-1}, and the system's maximum frequency variation rate is set to ±1 Hz/s. The system inertia margin diagram is obtained, as shown in Figure 3(a). Figure 3(b) presents a top view of Figure 3(a) as observed along the z-axis onto the *oy*-plane.

doi:10.1088/1742-6596/3031/1/012012

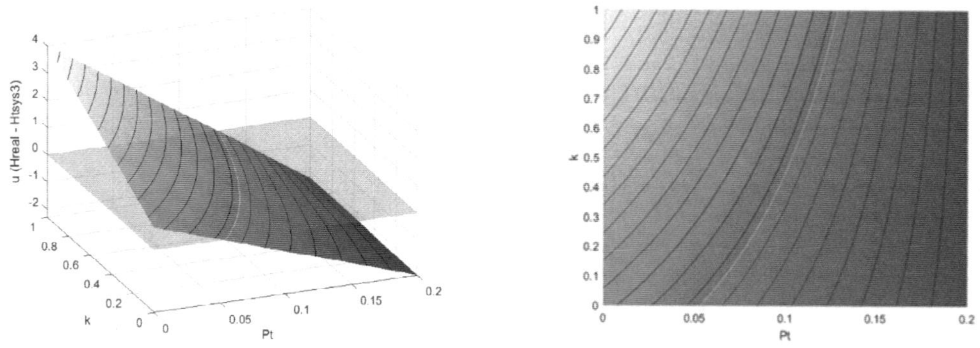

(a) (b)

Figure 3. System inertia margin diagram.

In Figure 3: k represents the thermal power proportion, and Pt denotes the ratio of power imbalance to system load.

3. Simulation

A three-machine nine-bus system is established for power system simulation. The simulation system includes two thermal generating units, each with a rated capacity of 100 MW, and a wind farm with an installed capacity of 52 MW. The wind farm consists of 35 DFIGs, each with a rated power of 1.5 MW. The structural and operational parameters of a single wind turbine are as follows: rated power of 1.5 MW; rated wind speed of 13 ms^{-1}; rotor radius of 40 m; at a pitch angle of 0°, the optimal tip-speed ratio λ_{opt} is 6.3249, corresponding to a maximum power coefficient $C_{p\max}$ of 0.4382; the rated rotor angular velocity is 2.056 rad/s, corresponding to the synchronous speed; the safe rotor angular velocity range is 0.7–1.2 p.u.; the maximum allowable electromagnetic torque $T_{e\max}$ is 1.35 p.u.; the grid-side inverter capacity limit is 0.25 p.u.; and the initial reactive power at the grid side is 0.03 p.u. It is worth noting that the per-unit base for wind farm power is its rated power. The air density ρ is 1.295 kg/m³.

The wind turbines participate in grid frequency regulation through integrated inertia control, with control parameters K_{pf}=0.4672, K_{df}=1.7514. The system simulation diagram is shown in Figure 4.

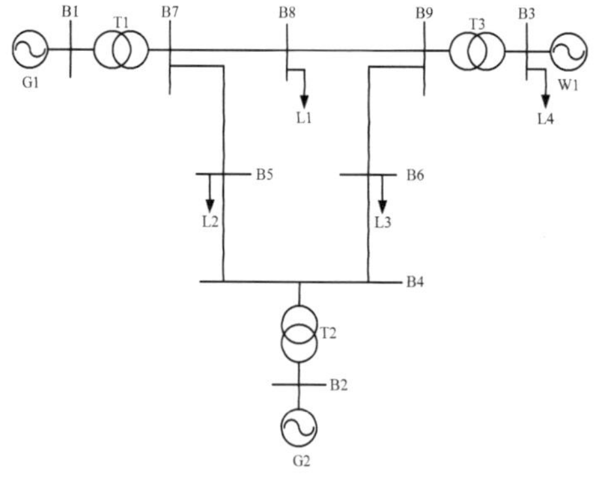

Figure 4. System simulation diagram.

3.1. Impact of wind speed prediction errors on wind turbine inertia

To intuitively and accurately observe the impact of wind speed prediction errors on wind turbine inertia, this section takes a forecasted wind speed of 12 ms^{-1} at a specific time as an example. Using historical forecast data and actual wind speed errors, 100 random samples are extracted through Latin hypercube sampling. Based on the sampled data shown in Figure 5, it is observed that the actual wind speed primarily falls within the range of [11.73, 12.18] ms^{-1}. Confidence level analysis of the data yields the results shown in Table 1, with the confidence level for $H_{t,W}$ set to 97%.

Table 1 Wind turbine inertia considering prediction error.

v_p (ms)	ω (p.u.)	P_e (p.u.)	H_W (p.u.)	$H_{t,W}$ (p.u.)
7	0.64	0.1561	0.1788	0.1446
8	0.73	0.2330	0.2632	0.2339
9	0.82	0.3318	0.3848	0.3587
10	0.91	0.4552	0.5254	0.4866
11	1	0.6058	0.6832	0.6361
12	1.09	0.7865	0.8578	0.8324
13	1.18	1	1	0.9737

The table results confirm the difference between the wind turbine inertia.

H_W under different forecasted wind speeds without considering prediction errors and the wind turbine inertia $H_{t,W}$ when prediction errors are considered.

3.2. Critical thermal power generation proportion considering wind speed prediction errors

Considering the operation of wind turbines in the MPPT region with supplemental inertia control, with $V_W = 0.75V_{Wn}$ and t_{end} =0.5s, it is derived from Equations (3) and (6) that H_W =0.83s. Setting $P_t = 5\%S_R$, Equation (9) yields $k_{f\lim} = -2.02\%$, indicating that the system can rely entirely on wind turbines to provide inertia and satisfy the RoCoF constraint.When $P_t = 7\%S_R$, the result is $k_{f\lim} = 12.56\%$, meaning the system requires at least 12.56% thermal power units. For $P_t = 10\%S_R$, $k_{f\lim} = 44.90\%$ .Without considering wind speed prediction errors, the corresponding $k_{f\lim}$ values are 4.89%, 9.72%, and 42.57%, respectively.

When the power imbalance during a disturbance is set to 7% of the system load, the derived critical proportion of thermal power is 12.56%. Simulations are conducted to validate the system under different proportions of thermal power, and the results are shown in Figure 6. PoTP represents the Proportion of Thermal Power.

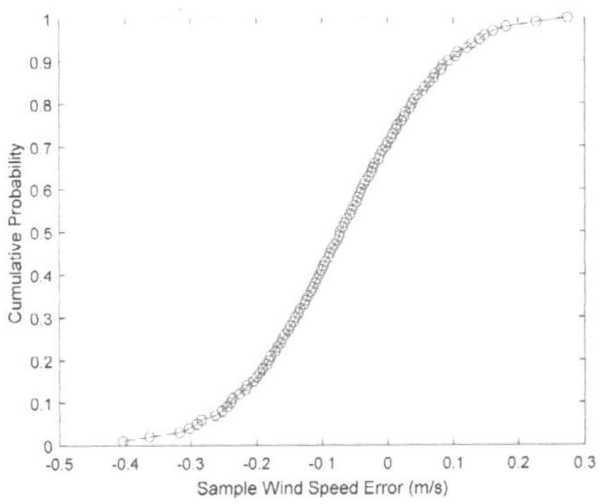

Figure 5. Latin hypercube sampling results considering wind speed prediction error.

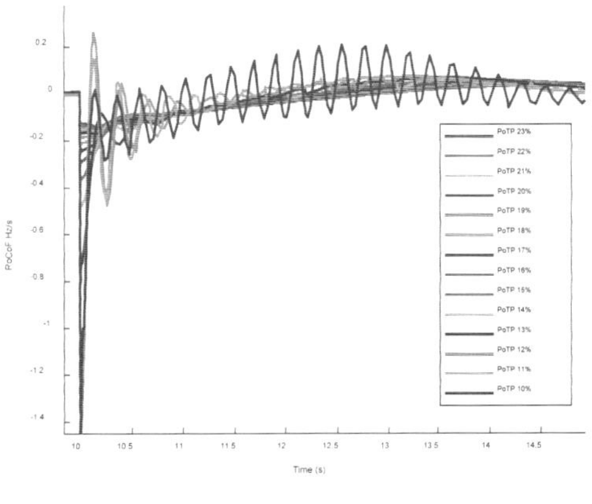

Figure 6. System frequency change rate when a disturbance occurs.

From the simulation results in Figure 6, it can be observed that when the thermal power proportion is 13%, the system's RoCoF reaches 0.817 Hz/s immediately after the disturbance, remaining within the safe threshold of 1 Hz/s. However, when the thermal power proportion is 12%, the system's RoCoF significantly exceeds 1 Hz/s at the moment of the disturbance. It can thus be concluded that the critical thermal power proportion lies between 12% and 13%. Therefore, under the boundary conditions provided in Section 3.3, the conclusion of a 12.56% critical thermal power proportion is validated.

Additionally, the 9.72% critical thermal power proportion, derived without considering wind speed prediction errors, poses a systematic instability risk. As shown in the PoTP 10% case in Figure 6, the system becomes unstable. This highlights the necessity of accounting for wind speed prediction errors.

4. Conclusion

Based on the aforementioned theoretical derivations and simulation results, the following key conclusions are drawn:

(1) This paper proposes a method for evaluating wind turbine inertia considering wind speed prediction errors. Simulations demonstrate the impact of wind speed prediction errors on wind turbine inertia, which increases the system's demand for the critical proportion of thermal power.

(2) The influence of wind speed prediction errors on the critical thermal power proportion can be reflected through the system's RoCoF. Simulation results validate the accuracy of the proposed critical thermal power proportion evaluation model.

(3) The critical proportion of thermal power units is proposed from the perspective of inertia based on the RoCoF constraint. Compared with the system's critical inertia metric, this approach is more intuitive and facilitates system operators in monitoring and maintaining the safety and stability of the power system.

Acknowledgments

This work was financially supported by the National Key R&D Program of China (2022YFB2403204)

References

[1] Ma N, Xie X, Li H and Zhao J (2024) Analysis of Inertia Demand for Renewable Energy Units Considering Dynamic Frequency Distribution. J. *Proceedings of the CSEE*, vol. 44, no. 09, pp. 3500-3508.

[2] Soni N, Doolla S and Chandorkar M C (2016) Inertia Design Methods for Islanded Microgrids Having Static and Rotating Energy Sources. J. *IEEE Transactions on Industry Applications*, vol. 52, no. 6, pp. 5165-5174.

[3] Ashabani S M and Mohamed Y A-R I (2012) A Flexible Control Strategy for Grid-Connected and Islanded Microgrids With Enhanced Stability Using Nonlinear Microgrid Stabilizer. J. *IEEE Transactions on Smart Grid*, vol. 3, no. 3, pp. 1291-1301.

[4] Imane B, Zakarya O and Ahmed A (2023) Frequency and ROCOF estimation under steady-state and dynamic conditions for three-phase grid-connected converters. J. *Electric Power Systems Research, Volume* 224,2023,109723,ISSN 0378-7796.

[5] Tu G, Li Y and Xiang J (2022) Coordinated Rotor Speed and Pitch Angle Control of Wind Turbines for Accurate and Efficient Frequency Response. J. *IEEE Transactions on Power Systems*, vol. 37, no. 5, pp. 3566-3576

[6] Pourbeik P et al. (2024) A Generic Model for Inertia-Based Fast Frequency Response of Wind Turbines and Other Positive-Sequence Dynamic Models for Renewable Energy Systems J. *IEEE Transactions on Energy Conversion*, vol. 39, no. 1, pp. 425-434

[7] Wang Z and Wu W (2018) Coordinated Control Method for DFIG-Based Wind Farm to Provide Primary Frequency Regulation Service J. *IEEE Transactions on Power Systems*, vol. 33, no. 3, pp. 2644-2659

[8] Björk J, Pombo D V and Johansson K H (2022) Variable-Speed Wind Turbine Control Designed for Coordinated Fast Frequency Reserves J. *IEEE Transactions on Power Systems*, vol. 37, no. 2, pp. 1471-1481.

[9] Zhu D, Guo X, Tang B, Hu J, Zou X and Kang Y (2024) Feedforward Frequency Deviation Control in PLL for Fast Inertial Response of DFIG-Based Wind Turbines J. *IEEE Transactions on Power Electronics*, vol. 39, no. 1, pp. 664-676

[10] Yan K, Li G, Zhang R, Xu Y, Jiang T and Li X (2024) Frequency Control and Optimal Operation of Low-Inertia Power Systems With HVDC and Renewable Energy: A Review J. *IEEE Transactions on Power Systems*, vol. 39, no. 2, pp. 4279-4295

[11] Wen Y and Lin X (2021) Assessment of Minimum Inertia Requirement for Microgrids under Islanded and Grid-connected Modes J. *Proceedings of the CSEE*, vol. 41, no. 6, pp. 2040-2053

Coupling response of offshore floating wind turbine-wharf in process of installation

Xinghua Shi [1,*], Yiming Wang [1], Honglong Li [2], Chunhan Wu [2]

[1] School of Naval Architecture and Ocean Engineering, Jiangsu University of Science and Technology, Zhenjiang 212003, China
[2] Zhongtian Technology Marine Engineering Co., Ltd., Nantong 226016, China

*Corresponding author's e-mail: xinghuashi@just.edu.cn

Abstract. With the development of offshore wind power construction, high construction and maintenance costs have become a key issue restricting its development. The installation process is the key in construction. Therefore, based on AQWA and Orca Flex, a coupled numerical model of a floating wind turbine-wharf was established to study the response of the floating platform, the installation docking position, cables, and fenders in the floating installation process of the offshore floating wind turbine, and optimize the installation process. The results show that the hydrodynamic performance of the floating wind turbine is changed after coupling with the wharf, and cannot be ignored in the installation and maintenance. With the tower installation, the motion and acceleration of the installation docking position increase, and the response can be controlled by configuring the balance weight rationally.

1. Introduction

Since the installation of the first floating wind turbine Hywind Demo in 2009, the global floating wind power is developing rapidly. By the end of 2022, the global installed capacity of offshore wind power has reached 64.31 GW [1]. The construction of demonstration projects provides good conditions for the research on floating wind turbine technology., such as Hywind Scotland at Northeast Sea of Scotland built in 2017 and scheduled to be decommissioned in 2022 after five years of work [2].

The floating platform is made of steel or concrete, it can complete the assembly of the platform, tower, and wind turbine in the port terminal. The whole towing will be built after lifting the wind turbine in the terminal [3], and it is the most feasible construction scheme. This construction method has been adopted for the floating wind power platform built by the China National Offshore Oil Corporation [4], the "wind power-fishery integration" marine ranching of Longyuan Power [5], the WindFloat I project built by EDP and Principle Power [6], and the project of Floatgen in France [7].

In order to realize the large-scale and commercial development of floating wind turbines, the key technical problems that need to be overcome include floating platforms, mooring systems, integrated simulation, water tank experiments, installation technology, etc. [3]. Among them, installation accounts for floating wind turbines for a large proportion of the total investment. In bottom-fixed and floating projects, the installation accounts for 6.3 % and 2.4 % of the total [8]. Fang Z B and Sun Z G [9] used a numerical model to calculate the cable tension under various environmental conditions, found that the cable tension is large when the mooring is short, and proposed a wharf mooring scheme. Min Y [10] used AQWA to study the wharf mooring scheme and motion response of the floating platform in an installation wharf and found that a higher mooring position is beneficial. However, they did not pay attention to the motion response of installation docking positions, which is important during the installation process.

For this problem, INO WINDMOOR, a semi-submersible floating wind turbine platform designed by Carlos et al. [11], is taken as the research object in this paper. The motion of, acceleration, mooring line tension, and fender force of the floating platform and installation docking positions during the construction and installation of the wharf under environmental loads were studied. And a reasonable installation plan was proposed to improve installation safety.

2. Numerical model of a floating wind turbine in a wharf

2.1. The parameters of floating wind turbines and wharf

Semi-submersible floating wind turbine INO WINDMOOR, equipped with 12WM turbine. The main dimensions and inertia properties are shown in Table 1. The design water depth of floating wind turbine work is 150 m. Considering that the water depth of the wharf is usually shallower than the floating wind turbine's work depth, the shallow water effect will cause the heave response of the floating platform to increase, and increase the risk of structural bottom contact. Therefore, it is necessary to study the response in the wharf environment. The main dimensions of the wharf are shown in Table 2.

Table 1. The main dimensions and inertia properties. The radii of the gyration refer to the hull's center of gravity.

Property	Value	Property	Value
Column diameter (m)	15.0	Total substructure mass (t)	11974.0
Column height (m)	31.0	Total substructure CG_x (m)	-5.91
Pontoon width (m)	10.0	Total substructure CG_y (m)	0.0
Pontoon height (m)	4.0	Total substructure CG_z (m)	-9.7
Center-center distance (m)	61.0	Total substructure R_{xx} (m)	23.66
Deck beam width (m)	3.5	Total substructure R_{yy} (m)	18.63
Deck beam height (m)	3.5	Total substructure R_{zz} (m)	28.10
Full floating wind turbine displacement (t)	14176.1	Full floating wind turbine draft(t)	15.5

Table 2. The main dimensions of a wharf. The design water depth of the wharf apron is the basal level.

Property	Value	Property	Value
Total length (m)	426.0	Total width (m)	37.0
Top elevation (m)	+8.0	Wharf apron water depth (m)	26.0
Highest water level (m)	+2.0	Lowest water level (m)	-1.5

In order to facilitate the installation operation, the nacelle and tower on the floating platform are close to the side of the wharf. The local system position of the floating platform refers to a right-handed Earth-fixed coordinate system. The X-Y plane coincides with the mean water level, and the Z-axis is positive upwards. The wharf coordinate system is defined as follows: the floating platform center is the origin of the coordinate, the X_w-axis is parallel to the wharf, the right direction of the wharf is the positive direction of the coordinate axis, the normal direction of the wharf is the Y_w-axis direction, the distance for from the wharf is the positive direction and the Z-axis is perpendicular to the waterline, being the same as floating platform's position, as shown Figure 1(b) and (c).

2.2. The wharf mooring scheme

Ten bollards are arranged on the wharf, and mooring points are arranged on the floating platform. The floating platform is restrained by connecting the cable between the mooring point and the bollards, to keep it stable during the installation process. Because the gravity center of the floating wind turbine is high, the motion and acceleration response at the docking point should be reasonably controlled during the installation process. To avoid a great change in the docking position during the installation process, 24 mooring points are arranged on the floating platform. Each mooring point has a cable to connect the bollards, and double cables are arranged at the key position to enhance its safety. Each bollard is subjected to the resultant force of the tension of the cable connected to it within its safe range. The cable arrangement is shown in the figure. The fender adopts a TDA2000 nonlinear fender, and the positions are shown in Figure 1(c).

2.3. Validation of the numerical method

Maxime et al. [12] made the 1:40 scale according to the INO WINDMOOR designed by Carlos et al. [11]. The experimental model is similar to the prototype in geometry and gravity. The model takes the centroid of the floating platform as the origin of the coordinate system. The error of the center of gravity of the experimental model relative to the prototype is less than 2 %, and the error of the moment of inertia is less than 3 %. The test results are reliable. The experimental model is shown in Figure 1(a).

(a) Experimental model [12] (b) Numerical model (c) Coupled floating wind turbine-wharf

Figure 1. The INO WINDMOOR model of model test and simulation.

The INO WINDMOOR proportional numerical model is established by OrcaFlex software. The self-built model is placed in still water to make the model decay freely under still water conditions. The simulation time is set to 10800 s, and the free decay curve of the model is obtained. The 6-degree-of-freedom power spectral density (PSD) of the platform is obtained by the Fast Fourier transform of the attenuation results obtained by the model time history simulation. The peak of the power spectral density is the natural frequency, and the value is shown in Table 3. The experimental model is compared with the numerical model in this paper. The maximum error of the numerical model relative to the experimental model is 4.39 %. Therefore, the calculation results of the numerical model are reliable.

Table 3. The natural period of experimental model and numerical model.

	Experimental model	Numerical model
Surge natural period [s]	93.7	93.98
Heave natural period [s]	16.4	16.43
Pitch natural period [s]	30.3	28.97

3. Coupling analysis of floating wind turbine-wharf in the installation process

3.1. Hydrodynamic parameters of frequency domain about coupled floating wind turbine-wharf

The wave is usually perpendicular to the wharf in practice, so the wave is along the negative Y_w-axis in this paper.

3.1.1. RAO (Response Amplitude Operator). The difference in RAO is small in the translational direction. The RAO response of the wind turbine under the coupling condition of the wharf is different from that without coupling in rotation, as shown in Figure 2. There is a great difference between the floating platform and INO WINDMOOR. The difference is minor after the balance weight is added to the floating platform.

3.1.2. First-order wave forces. It can be seen from Figure 3 that the first-order wave force changes greatly after the INO WINDMOOR is coupled with the wharf. Especially in the higher frequency band, the wave force in the surge and sway increases, the heave decreases, and the yaw direction decreases greatly after coupling with the wharf. Compared with the floating platform, the first-order wave force of INO WINDMOOR decreases greatly in heave, pitch, roll, and yaw, and the wave force decreases less in the surge and heave. After the balance weight is installed on the platform, the first-order wave force of the platform is similar to that of INO WINDMOOR in the surge, sway, heave, and yaw. In the pitch and roll, the first-order wave force changes with the wave frequency. The trend is close to that of the platform without any balance weight, but the maximum first-order wave force does not exceed that of the installed INO WINDMOOR.

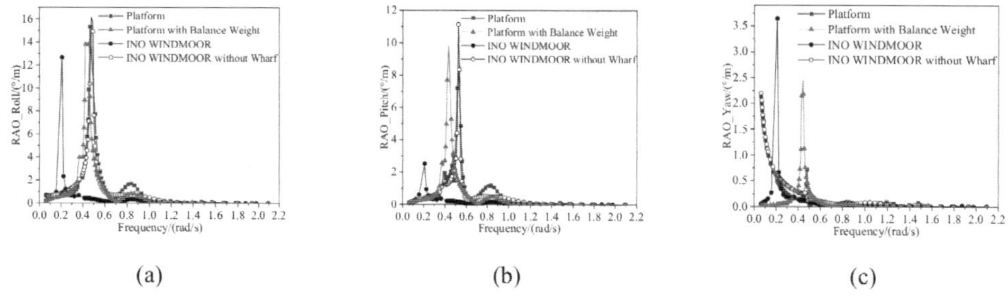

Figure 2. RAOs.

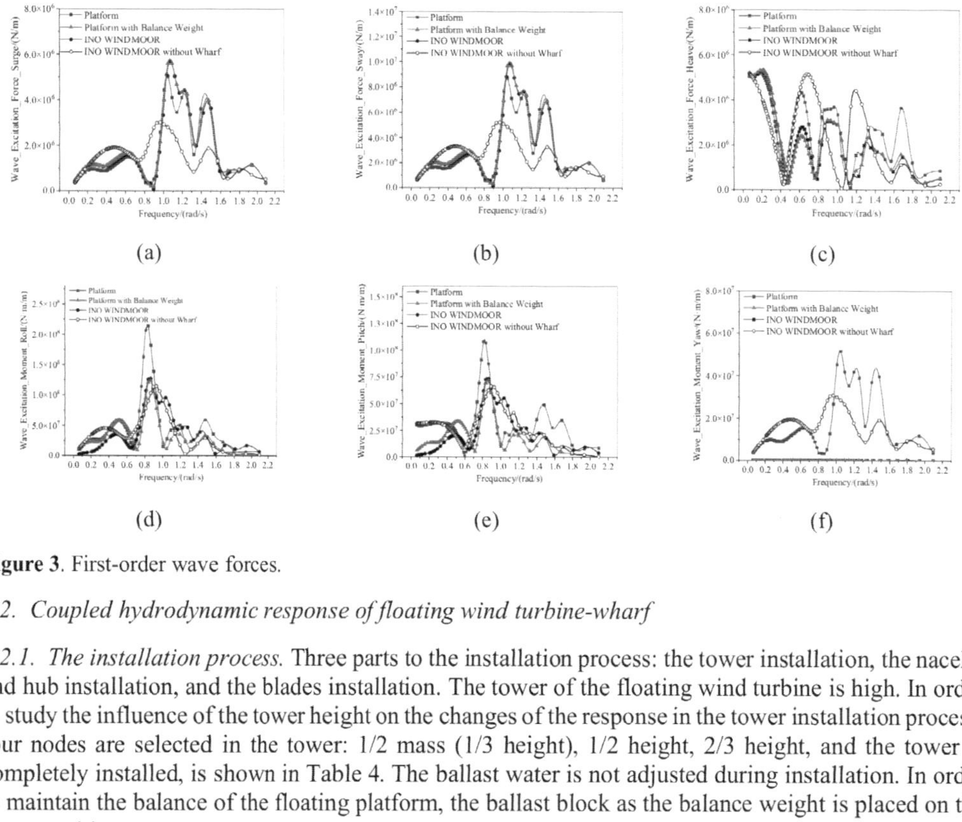

Figure 3. First-order wave forces.

3.2. Coupled hydrodynamic response of floating wind turbine-wharf

3.2.1. The installation process. Three parts to the installation process: the tower installation, the nacelle and hub installation, and the blades installation. The tower of the floating wind turbine is high. In order to study the influence of the tower height on the changes of the response in the tower installation process, four nodes are selected in the tower: 1/2 mass (1/3 height), 1/2 height, 2/3 height, and the tower is completely installed, is shown in Table 4. The ballast water is not adjusted during installation. In order to maintain the balance of the floating platform, the ballast block as the balance weight is placed on the tower position.

The pre-tension of the cable is 300 - 320 kN (6.7 % breaking tension). The cables are not adjusted during the installation process. The balance weight is a hollow cylinder 15 m in outer diameter, 11 m in inner diameter, and 3.2 m in height. The hollow cylinder (material steel with a density 7.86 t/m^3) is taken on the tower located on the platform. Plan-A is to dynamically adjust the overall weight by removing part of the balance weight, and keeping the mass unchanged during installation; plan-B is to remove all the balance weight after the cables are fixed; Plan-C is to keep the balance weight on the platform during the whole installation process until completion; plan-D is to remove half of the balance weight after the cables after fixed, and remove the remaining part of the balance weight after the installation is completed.

Table 4. The installation stage.

Stage	Composition	Stage	Composition
I	Floating platform	V	Floating platform+ Tower (All)
II	Floating platform+ Tower (1/2 tower mass)	VI	Floating platform+ Tower+ Nacelle
III	Floating platform+ Tower (1/2 tower height)	VII	Floating wind turbine INO MINDMOOR
IV	Floating platform+ Tower (2/3 tower height)		

3.2.2. The motion response of the floating platform. As shown in Figure 4, under different installation balance weight schemes, it can be seen from the motion amplitude response of the floating platform installation process that under Plan-A, as the balance weight scheme is dynamically adjusted, the motion amplitude response of X_w, Z, RY_w, and RZ changes smoothly in the installation process. The motion amplitude response of Y_w and RX_w decreases gradually with the completion of installation. Plan-B without any balance weight and Plan-C with the whole balance weight are more unstable than Plan-A, the motion response is weak in some installation stages, but it suddenly increases at a certain stage, with high safety risks. Under Plan-D with only half of the balance weight, the motion amplitude response of the floating platform is the same as that of Plan-A in the Y_w and RX_w in the installation process. Other direction motion responses are between Plan-B and Plan-C. The motion response amplitude of each direction in Plan-D does not exceed the maximum value of all schemes compared with other schemes.

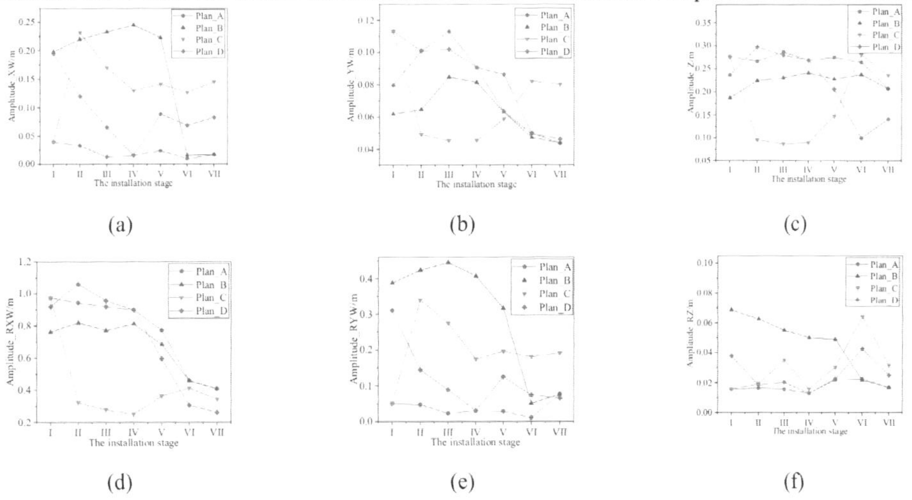

Figure 4. The motion amplitude response of the floating platform.

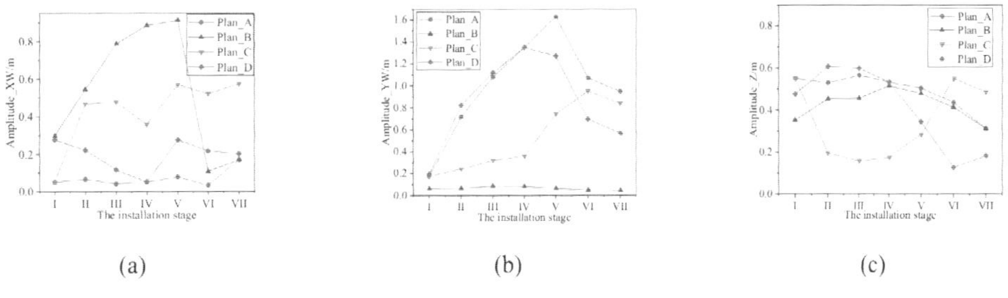

Figure 5. The motion response of installation docking positions.

3.2.3. The motion response of installation docking positions. The rotation motion amplitude response of the installation docking position is the same as the floating platform, but the translation motion response is different in the installation process. The motion amplitude response of the installation docking position is several times of the floating platform. With the installation, the difference increases between the motion amplitude response of the installation docking position and the platform, and the heavier balance weight can control the motion response of the installation position,is shown in Figure 5.

3.2.4. The acceleration of installation docking positions, maximum mooring tension and maximum fender force. As shown in Figure 6, the maximum acceleration response of the installation docking position increases with the installation. During the installation process, the maximum tension of the cable and the fender force can be controlled by the balance weight. However, the excessive balance weight has a large fender force at the end of the installation.

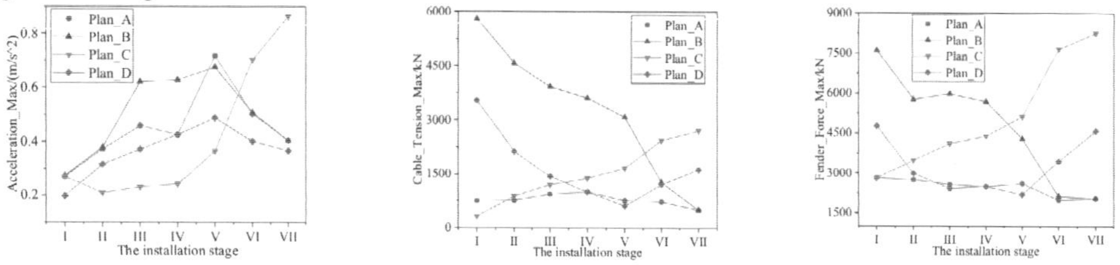

(a) (b) (c)

Figure 6. The acceleration of installation docking positions, maximum mooring tension, and maximum fender force.

4. Conclusions

Based on the results and discussions presented above, the conclusions are obtained as below:

(1) It is shown that the first-order wave force of INO WINDMOOR is different after being coupled with the wharf and also different from the platform before installation. Adding the balance weight can effectively control this difference.

(2) As installation proceeds, the motion amplitude of the installation docking positions increases rapidly with the increase of the tower height, being several times that of its platform. The balance weight is very helpful to control the motion at the installation docking positions.

(3) A large balance weight will lead to a large fender force at the end of the installation, making it necessary to allocate the balance weight reasonably.

Acknowledgments

This work was financially supported by the Key Research and Development Program for Core Technology by the Nantong City government (JB2024009).

References

[1] GWEC (2023) Global Offshore Wind Report 2022. https://gwec.net.
[2] Equinor (2022) Hywind Scotland Pilot Park Decommissioning Programme. https://www.equinor.com/energy/hywind-scotland.
[3] Jiang Z Y (2021) Installation of offshore wind turbines: A technical review. *Renewable and Sustainable Energy Reviews*. Volume 139, April 2021, 110576.
[4] Xinhua (2023) Construction milestone for China's first deep-sea floating wind power platform. http://www.china.org.cn/china/Off_the_Wire/2023-01/03/content_85038453.htm.
[5] CHN Energy (2023) CHN Energy launched its first "offshore wind power plus" project. https://www.ceic.com/gjnyjtwwEn/xwzx/202305/4c7b4bc72666430fad69269aa9f51b56.shtml.
[6] Principle Power (2024) WindFloat1. https://www.principlepower.com/projects/windfloat1.

[7] CENTRALE NANTES (2017) Floatgen, first floating-wind turbine installed off the French coast. https://research.ec-nantes.fr/en/industrial-chairs-and-joint-laboratories/floatgen-first-floating-wind-turbine-installed-off-the-french-coast.
[8] Stehly TJ, Beiter PC, Heimiller DM, Scott GN (2018) 2017 Cost of wind energy review. Technical Report, NREL/TP-6A20-72167. Denver, CO, USA: NREL; 2018.
[9] Fang Z B, Sun Z G (2022) Wharf Mooring Force Analysis of Floating Wind Power Platform. *JIANGSU SHIP (02)*, pp.1–3+36+60.
[10] Min Y (2023) Wharf Mooring Scheme and Dynamic Response of Offshore Floating Wind Turbine. *CHINA OFFSHORE PLATORM (04)*, pp. 36–41.
[11] C. E. Silva de Souza, P. A. Berthelsen, L. Eliassen, et al. (2021) Definition of the INO WINDMOOR 12 MW base case floating wind turbine. SINTEF Ocean, OC2020 A-044.
[12] M. Thys, C. E. Souza, T. Sauder, et al. (2021) Experimental Investigation of the Coupling Between Aero- and Hydrodynamical Loads on A 12 MW Semi-Submersible Floating Wind Turbine, in Proceedings of the International Conference on Offshore Mechanics and Arctic Engineering-OMAE.

Journal of Physics: Conference Series **3031** (2025) 012014 doi:10.1088/1742-6596/3031/1/012014

Multidimensional industry growth trend analysis and business cycle forecasting model based on power big data

Zhiqiang Lan[1*], Yongjie Guo[1], Linling Mao[2], Yanan Yu[2], Yajing Chen[3]

[1] State Grid Fujian Marketing Service Center, Fuzhou, Fujian, 361000, China

[2] Fujian Power Supply Service Co., Ltd. Xiamen, Fujian, 361000, China

[3] State Grid Fujian Electric Power Co., Ltd., Quanzhou Power Supply Company, Quanzhou, Fujian, 362000, China

*Corresponding author's e-mail: mattlan0228@outlook.com

Abstract. With the continuous development of the power industry, accurately predicting its growth trends and business cycle fluctuations has become a crucial task for optimizing resource allocation. Traditional forecasting methods often rely on linear models, which struggle to capture complex nonlinear relationships and large-scale data. This paper proposes a Deep Graph Convolutional Network (D-GCN)-based model for analyzing the growth trends and forecasting the business cycle of the power industry. By integrating graph neural networks (GNNs), deep graph convolution, and adaptive graph convolution techniques, the model captures the multidimensional economic and technical interconnections within the power industry. Through the processing of multi-source heterogeneous data and the fusion of spatiotemporal features, the model enhances prediction accuracy and robustness. Furthermore, by incorporating deep causal inference and deep reinforcement learning, the forecasting accuracy and interpretability of the business cycle prediction are further improved. Experimental results demonstrate that the proposed model effectively predicts the growth trends of the power industry, providing valuable insights for industry decision-making.

1. Introduction

As the power industry continues to develop and the economic environment changes, accurately predicting its growth trends [1] and business cycle fluctuations [2] has become crucial for scientific decision-making and optimizing resource allocation. Traditional forecasting methods in the power industry often rely on linear models and time series analysis [3]. However, these approaches struggle to provide accurate predictions and deep insights, particularly when faced with complex nonlinear relationships and large-scale multidimensional data [4]. In particular, under dynamic policy environments and fluctuating energy demands, the limitations of traditional methods have become increasingly evident. Therefore, improving the accuracy and adaptability of business cycle forecasting in the power industry using advanced technologies has become an urgent research challenge.

Currently, deep learning [5] and graph neural networks (GNNs) [6] have achieved significant success in forecasting and analysis across various industries, yet their application in the power industry remains in an exploratory stage. Existing studies predominantly focus on time-series-based forecasting [7], overlooking the complex interconnections between regions, policies, and markets. This oversight limits model flexibility and applicability. Moreover, many traditional models fail to extract underlying causal relationships within the data, preventing them from fully capturing the dynamic changes in the power industry. As a result, current methods exhibit limitations in accuracy, robustness, and interpretability, highlighting the need for technological breakthroughs [8].

To address these challenges, this paper proposes a Deep Graph Convolutional Network (D-GCN)-based model for growth trend analysis and business cycle forecasting in the power industry. By incorporating graph neural networks, deep graph convolution, and adaptive graph convolution

Journal of Physics: Conference Series **3031** (2025) 012014 doi:10.1088/1742-6596/3031/1/012014

techniques, the model effectively captures the complex economic and technical interconnections among power industry entities. Additionally, by integrating spatiotemporal feature analysis, the model enhances prediction accuracy and robustness. Through a comprehensive application of various deep learning techniques, the proposed approach not only overcomes the limitations of existing methods but also provides new insights and solutions for business cycle analysis and forward-looking forecasting in the power industry.

2. Deep learning-based growth trend analysis and business cycle forecasting model for the power industry

2.1. Data collection and preprocessing

To ensure the accuracy and applicability of the proposed growth trend analysis and business cycle forecasting model for the power industry, this study collects multi-dimensional data, including power consumption, industry production data, policy adjustments, and market demand. The primary data sources are the State Grid Corporation, power dispatch centers, government statistical departments, and industry research institutions. These datasets cover different time periods and regions, forming a broad spatiotemporal information system. During data collection, raw data is obtained through API interfaces, web scraping, and publicly available databases to ensure timeliness and completeness. Additionally, to enhance the model's generalization ability, socio-economic data is incorporated to provide a comprehensive evaluation of the power industry's growth trends.

The collected data undergoes cleaning to remove missing values and anomalies, and data from different sources is aligned to ensure consistency. For time-series data, missing values are filled using interpolation methods, and time-step features are constructed using the sliding window approach. To handle multi-source heterogeneous data, normalization and standardization techniques are applied, allowing data with different units to be analyzed on a unified scale. Additionally, given that some data exhibits periodic or trending characteristics, wavelet transform or exponential smoothing is used for detrending. All processed data is then formatted to fit the input structure of deep learning models, laying the foundation for subsequent modeling and analysis.

2.2. Growth trend analysis model for the power industry based on D-GCN

This study proposes a model based on Deep Graph Convolutional Networks (D-GCN) to comprehensively analyze growth trends in the power industry and enable effective forecasting. The model integrates Graph Neural Networks (GNN), Deep Graph Convolutional Networks, and Adaptive Graph Convolutional Networks (AGCN) to fully capture the economic interconnections between industries, the dynamic evolution of time-series data, and the impact of spatial features, thereby providing more precise modeling capabilities for power industry growth trend analysis.

In this model, various entities within the power industry and their related regions are modeled as a graph structure. In the graph, nodes represent specific entities in the power industry or regions, and their features include power consumption, industrial output, and other key indicators. Edges represent economic or technological relationships between different nodes, with edge weights reflecting the strength of these relationships (e.g., supply chain connections or regional cooperation). To capture the complex interactions between different sectors, GNN propagates and updates node information based on the graph structure. Given a graph $G = (V, E)$, where V is the set of nodes and E is the set of edges, the node feature matrix is $X \in \mathbb{R}^{|V| \times d}$, where d is the dimensionality of each node's features. The core idea of GNN is to update each node's representation by aggregating feature information from neighboring nodes, effectively capturing inter-industry interactions.

To enhance the expressive power of the model, this study incorporates Deep Graph Convolutional Networks (D-GCN). By applying multi-layer graph convolution operations, D-GCN progressively fuses the features of neighboring nodes, enabling the model to capture more complex industry relationships and graph structural features. In traditional Graph Convolutional Networks (GCN), node features are updated using the following equation:

Journal of Physics: Conference Series **3031** (2025) 012014 doi:10.1088/1742-6596/3031/1/012014

$$H^{(l+1)} = \sigma\left(\hat{A}H^{(l)}W^{(l)}\right) \tag{1}$$

where $H^{(l)} \in \mathbb{R}^{|V|\times d_l}$ represents the node feature matrix at layer l, with dl being the feature dimension at that layer. $\hat{A}$ is the normalized adjacency matrix, which encodes the strength of relationships between adjacent nodes, defined as $\hat{A} = D^{-1/2}AD^{-1/2}$, where A is the adjacency matrix, and D is the degree matrix. $W^{(l)} \in \mathbb{R}^{d_l \times d_{l+1}}$ is the weight matrix of the l-th layer. σ is the activation function.

Through multi-layer graph convolution operations, D-GCN progressively aggregates features from neighboring nodes, thereby achieving deep modeling of complex industry relationships. By employing hierarchical feature extraction, D-GCN can effectively learn hidden interdependencies within the power industry, particularly in this highly interconnected and complex domain.

To further enhance the model's robustness and adaptability, an Adaptive Graph Convolutional Network (AGCN) module [9] is incorporated. Unlike traditional GCNs, AGCN dynamically adjusts the adjacency relationships between nodes based on their characteristics. This allows the model to flexibly adapt to changes in inter-node relationships. AGCN learns an adaptive adjacency matrix, ensuring that the graph structure evolves based on node features at each layer. The adjacency matrix is updated using the following equation:

$$\hat{A}^{(l)} = \mathrm{Softmax}\left(\frac{X^{(l)}\left(X^{(l)}\right)^T}{\sqrt{d_l}}\right) \tag{2}$$

where $X^{(l)}$ is the node feature matrix at layer l; d_l is the feature dimension of layer l. The Softmax function normalizes the adjacency matrix weights, ensuring that each node's adjacency relationships adjust dynamically based on its features.

This adaptive mechanism optimizes the information propagation process between nodes, thereby improving the model's adaptability and forecasting accuracy.

Analyzing the growth trends of the power industry requires considering both inter-industry interactions and spatiotemporal data. Therefore, the proposed model employs a spatiotemporal feature fusion approach, incorporating both time-series and spatial data in the learning process.

Temporal features are captured using D-GCN, which models the evolutionary dynamics of the power industry over time.

Spatial features are extracted using a Convolutional Neural Network (CNN), which identifies spatial economic and policy-related patterns. The fused model structure is expressed as follows:

$$F_t = \mathrm{D-GCN}(X, A) \tag{3}$$

$$F_s = \mathrm{CNN}\left(X_{spatial}\right) \tag{4}$$

$$F_{\mathrm{final}} = \mathrm{Fusion}(F_t, F_s) \tag{5}$$

where F_t represents the temporal features extracted by D-GcN, reflecting the industry's development trends over time. F_s represents the spatial features extracted by CNN, capturing the influence of regional economies and policies. F_{final} is the final fused representation, integrating multi-dimensional spatiotemporal information.

By incorporating spatiotemporal feature fusion, the model comprehensively considers the variations in the power industry across different time periods and regions. This enables more accurate growth trend analysis and forward-looking forecasts.

The D-GCN-based growth trend analysis model for the power industry effectively integrates Graph Neural Networks, multi-layer deep graph convolutions, and adaptive graph convolutions. This integration helps to model complex inter-industry relationships. By incorporating spatiotemporal feature fusion, the model enhances prediction accuracy and robustness, offering a novel approach for analyzing business cycles and forecasting future trends in the power industry.

2.3. Forward-looking forecast model based on deep causal inference and DRL

The economic climate of the power industry is influenced by multiple factors, such as policy adjustments, energy consumption, and market demand. Traditional time series forecasting methods often rely on correlation-based modeling while ignoring the underlying causal relationships between variables. This limitation makes it difficult to interpret complex economic behaviors and weakens the generalization ability of predictions. To enhance forecasting accuracy and interpretability, this study proposes a forward-looking forecast model that integrates Deep Causal Inference (DCI) with Deep Reinforcement Learning (DRL). The model first employs causal inference techniques to identify key driving factors and construct a causal relationship network. Then, reinforcement learning is used to optimize prediction strategies, allowing the model to adapt to dynamic market changes and improve decision-making robustness.

For causal inference, this study adopts the Structural Causal Model (SCM) [10] to construct a causal network for the power industry and establishes a causal graph using causal discovery algorithms. Given a variable set $X = \{x_1, x_2, \cdots, x_n\}$, where each variable xi may be influenced by other variables, SCM is represented by the following structural equation:

$$x_i = f_i(\mathrm{Pa}(x_i), \epsilon_i) \tag{6}$$

where $\mathrm{Pa}(x_i)$ denotes the direct causal variables of x_i, $f_i(\cdot)$ is a nonlinear function, and ϵ_i is an independent noise term.

To quantify causal effects, this study employs the Inverse Probability Weighting (IPW) method to compute the causal effect of a policy variable T on the industry climate index Y:

$$\mathbb{E}[Y|do(T=t)] = \sum_x \mathbb{E}[Y|T=t, X=x]\, P(X=x) \tag{7}$$

To enhance the modeling capability of causal inference, a Deep Neural Network (DNN) is introduced to estimate causal effects. The Causal Effect Network (CEN) is trained to predict the impact of variables, represented as:

$$\hat{Y} = f_\theta(X, T) \tag{8}$$

where $f_\theta(\cdot)$ is parameterized by a neural network and optimized through supervised learning, enabling the model to capture complex nonlinear causal relationships effectively.

For reinforcement learning, this study integrates Deep Q-Network (DQN) and Policy Gradient methods to optimize forecasting strategies. The reinforcement learning model follows a State-Action-Reward framework for adaptive learning. State S_t: Includes the current industry climate index, electricity consumption, market demand, etc. Action A_t: Represents different forecast adjustment strategies. Reward (R_t): Defined based on prediction errors or market responses, guiding the model toward optimization. The objective of reinforcement learning is to learn the optimal policy $\pi^*(A|S)$ to maximize the long-term return:

$$J(\theta) = \mathbb{E}\left[\sum_{t=0}^{T} \gamma^t R_t\right] \tag{9}$$

where γ is a discount factor controlling the importance of future rewards. The Q-learning process updates the Q-value using the following formula:

$$Q(S_t, A_t) = R_t + \gamma \max_{A'} Q(S_{t+1}, A') \tag{10}$$

Given the strong temporal dependencies in industry climate data, this study further integrates Long Short-Term Memory (LSTM) networks with reinforcement learning. The LSTM model controls information flow through gating mechanisms, with core update equations as follows:

$$\begin{gathered} f_t = \sigma\left(W_f[h_{t-1}, x_t] + b_f\right) \\ i_t = \sigma(W_i[h_{t-1}, x_t] + b_i) \\ C_t = f_t \odot C_{t-1} + i_t \odot tanh(W_c[h_{t-1}, x_t] + b_c) \\ o_t = \sigma(W_o[h_{t-1}, x_t] + b_o) \\ h_t = o_t \odot \tanh(C_t) \end{gathered} \tag{11}$$

where f_t, i_t, o_t represent the forget gate, input gate, and output gate, respectively, C_t is thememory cell, and h is the hidden state, The output of LSTM serves as the input for reinforcement learning states, allowing the agent to optimize decision-making based on long-term historical information, thereby improving prediction accuracy and adaptability.

The proposed forward-looking forecast model, integrating Deep Causal Inference and Deep Reinforcement Learning, enhances both the interpretability and adaptability of predictions:

Causal Inference Module: Constructs a causal network and employs deep learning to model nonlinear causal effects, improving the identification of key influencing factors in the power industry.

Reinforcement Learning Module: Combines DQN, Policy Gradient, and LSTM to optimize forecasting strategies dynamically in changing market conditions.

Experimental results demonstrate that the model outperforms traditional time series forecasting methods in terms of accuracy, interpretability, and adaptability, providing a powerful technical framework for forward-looking analysis and scientific decision-making in the power industry.

2.4. Model training

To ensure the effectiveness and generalization ability of the model, this study follows a systematic training process, including data preprocessing, training-validation set partitioning, hyperparameter optimization, loss function selection, and optimization strategies.

First, the dataset consists of time-series data from the power industry, including key variables such as the industry prosperity index, electricity consumption, policy changes, and market demand. The data partitioning follows the Rolling Window method [11], enabling the model to adapt to future data prediction needs. Given the total dataset D it is split according to a ratio r, where the training set D_{train} is used for model learning, and the validation set D_{val} is used for performance evaluation. The sliding window strategy is applied to enhance the model's ability to capture time-series patterns.

The model's hyperparameters significantly impact its predictive performance. Therefore, this study employs Bayesian Optimization to automatically tune hyperparameters, improving both training efficiency and prediction accuracy. The optimization objective function is defined as:

$$\theta^* = \arg\min_{\theta} \mathcal{L}(f_\theta(X), Y) \tag{12}$$

where $\mathscr{L}$ represents the loss function, and θ denotes the hyperparameters to be optimized including learning rateα, batch size B, number of LSTM hidden units h, and reinforcement learning discount factor γ. The optimization process first defines the hyperparameter search space and samples an initial set of parameter combinations. Then, a Gaussian Process (GP) is used to fit the loss function, and the next set of hyperparameters is selected based on the Expected Improvement (EI) criterion. This process iterates until the stopping condition is met.

For loss function selection, this study adopts a combination of Mean Squared Error (MSE) and Smooth L1 Loss to balance prediction accuracy and stability. The MSE loss is defined as:

$$\mathcal{L}_{\text{MSE}} = \frac{1}{N}\sum_{i=1}^{N}(y_i - \hat{y}_i)^2 \tag{13}$$

Smooth L1 Loss is introduced to mitigate the influence of outliers and is defined as:

$$\mathcal{L}_{\text{Smooth L1}} = \begin{cases} \frac{1}{2}(y_i - \hat{y}_i)^2, & |y_i - \hat{y}_i < 1| \\ |y_i - \hat{y}_i| - \frac{1}{2}, & otherwise \end{cases} \tag{14}$$

The final comprehensive loss function is formulated as:

$$\mathcal{L} = \lambda_1 \mathcal{L}_{\text{MSE}} + \lambda_2 \mathcal{L}_{\text{Smooth}}\text{L1} \tag{15}$$

where λ_1 and λ_2 are weight parameters, whose optimal values are determined experimentally.

For training optimization, this study employs the Adam optimizer for gradient updates, which accelerates convergence and avoids local optima. The rules for updating parameters are as follows:

$$m_t = \beta_1 m_{t-1} + (1 - \beta_1) g_t \tag{16}$$

$$v_t = \beta_2 v_{t-1} + (1 - \beta_2) g_t^2 \tag{17}$$

$$\widehat{m}_t = \frac{m_t}{1 - \beta_1^t}, \quad \hat{v}_t = \frac{v_t}{1 - \beta_2^t} \tag{18}$$

$$\theta_t = \theta_{t-1} - \alpha \frac{\widehat{m}_t}{\sqrt{\hat{v}_t} + \epsilon} \tag{19}$$

where g_t is the gradient, m_t and v_t are the first and second moment estimates of the gradient, and β_1 and β_2, are hyperparameters set to 0.9 and 0.999 by default. To prevent overfitting, this study incorporates Learning Rate Decay and Dropout strategies during training. The learning rate follows an exponential decay schedule, updated as follows:

$$\alpha_t = \alpha_0 \cdot \gamma^t \quad (20)$$

where α_0 is the initial learning rate, γ is the decay factor, and t is the current training epochAdditionally, Dropout is applied to LSTM and fully connected layers, randomly dropping neurons with probabilitypp to improve the model's generalization ability.

3. Results and analysis

3.1. Analysis of growth trends in the power industry

This study constructs an industry growth trend analysis network based on D-GCN to comprehensively evaluate growth trends in the power industry. It also integrates a spatiotemporal feature fusion model to explore key trends and influencing factors in depth. By analyzing the interactions between industries, this study identifies industries that play a core driving role in overall industrial development and reveals the spatiotemporal dynamics of industry growth, providing more scientific and forward-looking predictive support.

As shown in Figure 1, the power industry growth trend analysis network illustrates the complex relationships between different industries. In this network, nodes represent industries, while edges denote inter-industry influence weights, revealing the centrality and propagation effects of various industries. The study finds that the power industry is positioned at the center of the network. It plays a key driving role and has strong spillover effects on the manufacturing, finance, and technology industries. Meanwhile, some peripheral industries exhibit lower influence, indicating their relatively minor contribution to overall industry growth.

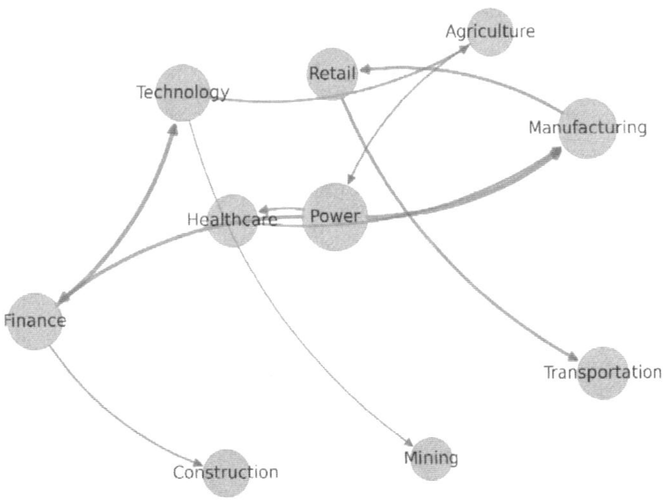

Figure 1. Power Industry Growth Trend Analysis Network.

As shown in Figure 2, this study analyzes the spatiotemporal patterns of power industry growth using the spatiotemporal feature fusion model. The results indicate that industry growth exhibits distinct cyclical fluctuations, driven by multiple factors such as market demand, policy adjustments, and technological innovations. Particularly at certain critical time points, industry development reaches inflection points, suggesting that external shocks (e.g., economic fluctuations or policy changes) significantly impact industry growth.

Journal of Physics: Conference Series **3031** (2025) 012014
doi:10.1088/1742-6596/3031/1/012014

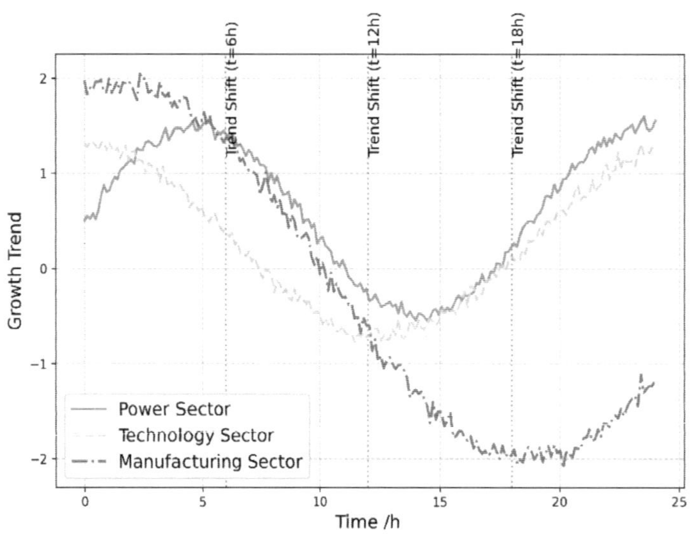

Figure 2. Spatiotemporal Pattern Analysis of the Power Industry.

Compared with traditional industry growth analysis methods, the proposed D-GCN and spatiotemporal feature fusion model offers significant advantages in the following aspects:

Graph neural networks (GNNs) efficiently model complex inter-industry dependencies, identifying key industries and their influence. Spatiotemporal feature fusion incorporates both time series and spatial correlations, capturing key trends and variation patterns in industry growth more accurately. The model fully utilizes multidimensional big data from the power sector, enhancing the accuracy of industry growth trend analysis and predictive capabilities. Therefore, the proposed model has extensive applications in growth analysis and business cycle forecasting for the power industry.

3.2. Forward-looking forecasting of industry prosperity

To evaluate the effectiveness of the Causal Inference and Deep Reinforcement Learning (CI-DRL) model in forward-looking industry prosperity forecasting, this study compares its performance with traditional forecasting methods in predicting the power industry prosperity index. The predictive accuracy of the model is validated by comparing forecasted values with actual values.

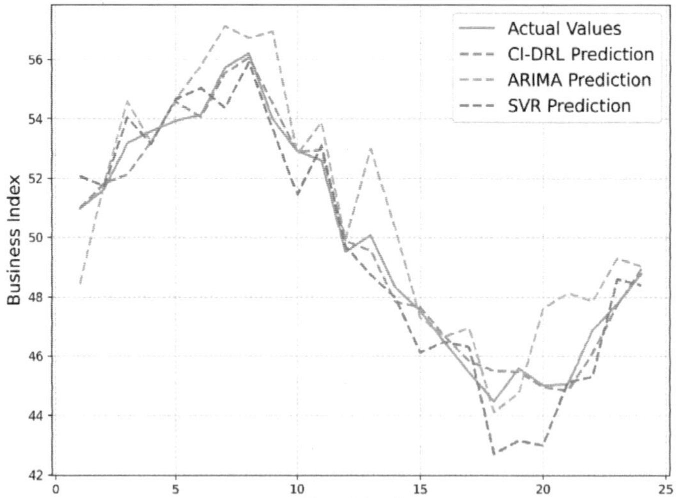

Figure 3. Comparison of Forward-Looking Forecasting Results.

As shown in Figure 3, the predictions made by the CI-DRL model closely align with the actual values, exhibiting lower prediction errors. This indicates that the CI-DRL model effectively captures fluctuations in the power industry prosperity index. In contrast, traditional models such as ARIMA and SVR show higher prediction errors, particularly in terms of accuracy and volatility. These results further validate the superior predictive accuracy of the CI-DRL model.

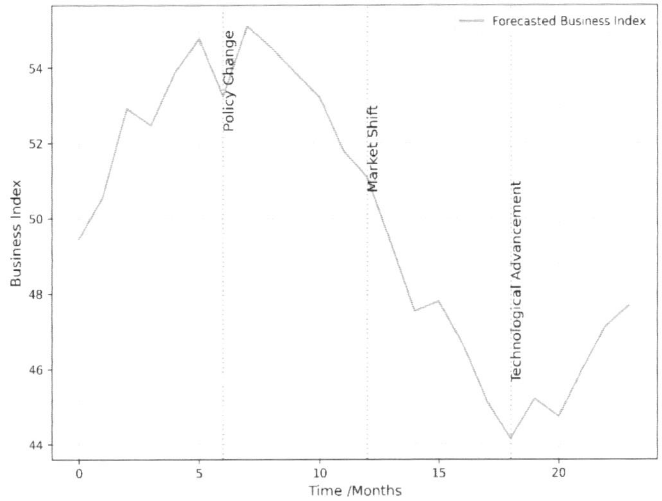

Figure 4. Forecasted Trends in Power Industry Prosperity.

Figure 4 illustrates the projected future trends of the power industry prosperity index and highlights three key factors: policy adjustments, market demand changes, and technological advancements. These factors play a crucial role in shaping future industry trends. They can also assist decision-makers in making appropriate predictions and strategic plans under various scenarios. The results indicate that as policies evolve, market conditions shift, and technological advancements progress, the power industry prosperity index shows a steady upward trend, suggesting a more positive growth outlook for the industry in the future.

The proposed CI-DRL model demonstrates significant advantages by accurately capturing the complex evolution of industry prosperity. Compared to traditional statistical models, the CI-DRL model leverages deep learning techniques to automatically identify hidden trends and causal relationships within data, providing more precise forecasting results. Its superiority is particularly evident in long-term trend prediction and complex scenario analysis.

4. Conclusion

This paper proposes and implements a deep learning-based model for analyzing growth trends and prosperity forecasting in the power industry, aiming to provide more accurate industry trend predictions and decision-making support. By collecting and preprocessing multi-dimensional data, the study constructs a comprehensive and high-quality dataset, incorporating power consumption, industry production data, policy adjustments, and socio-economic factors, laying a solid foundation for subsequent modeling and analysis. Based on this dataset, the proposed Deep Graph Convolutional Network (D-GCN) model integrates graph neural networks, deep convolution, and adaptive convolution techniques to effectively capture complex inter-industry relationships and enhance prediction accuracy through spatiotemporal feature fusion.

Furthermore, this study introduces a forward-looking prosperity forecasting model that combines deep causal inference and deep reinforcement learning. By constructing causal networks and integrating reinforcement learning strategies, the model improves its ability to identify key driving factors in the power industry and enhances its adaptability. Experimental results demonstrate that the proposed model outperforms traditional methods in terms of prediction accuracy, interpretability, and flexibility, providing stronger technical support for forward-looking analysis and scientific decision-making in the power industry. This research offers new perspectives and methodologies for industry prosperity analysis and future trend forecasting.

References

[1] Al-Shetwi A Q. Sustainable development of renewable energy integrated power sector: Trends, environmental impacts, and recent challenges[J]. Science of The Total Environment, 2022, 822: 153645.
[2] Jiang Z, Liu Z. Policies and exploitative and exploratory innovations of the wind power industry in China: The role of technological path dependence[J]. Technological Forecasting and Social Change, 2022, 177: 121519.
[3] Chen H B, Pei L L, Zhao Y F. Forecasting seasonal variations in electricity consumption and electricity usage efficiency of industrial sectors using a grey modeling approach[J]. Energy, 2021, 222: 119952.
[4] Klyuev R V, Morgoev I D, Morgoeva A D, et al. Methods of forecasting electric energy consumption: A literature review[J]. Energies, 2022, 15(23): 8919.
[5] Gao Y, Zhao T, Zheng Z, et al. Cotton leaf water potential prediction based on UAV visible light images and multi-source data[J]. Irrigation Science, 2025, 43(1): 121-134.
[6] Gao C, Zheng Y, Li N, et al. A survey of graph neural networks for recommender systems: Challenges, methods, and directions[J]. ACM Transactions on Recommender Systems, 2023, 1(1): 1-51.
[7] Wang D, Gan J, Mao J, et al. Forecasting power demand in China with a CNN-LSTM model including multimodal information[J]. Energy, 2023, 263: 126012.
[8] Brynjolfsson E, Jin W, McElheran K. The power of prediction: predictive analytics, workplace complements, and business performance[J]. Business Economics, 2021, 56: 217-239.
[9] Singh I P, Ghorbel E, Oyedotun O, et al. Multi-label image classification using adaptive graph convolutional networks (ML-AGCN)[C]//2022 IEEE International Conference on Image Processing (ICIP). IEEE, 2022: 1806-1810.
[10] Yang M, Liu F, Chen Z, et al. Causalvae: Disentangled representation learning via neural structural causal models[C]//Proceedings of the IEEE/CVF conference on computer vision and pattern recognition. 2021: 9593-9602.
[11] Jin Q, Tang C, Cai W. Research on dynamic path planning based on the fusion algorithm of improved ant colony optimization and rolling window method[J]. IEEE Access, 2021, 10: 28322-28332.

Journal of Physics: Conference Series 3031 (2025) 012015 doi:10.1088/1742-6596/3031/1/012015

Fault handling technology of new energy photovoltaic access in low-voltage power grid

Zhihua Xu [1,2,*], **Gang Ma** [1], **Song Hu** [2], **Jing Yu** [2], **Zhicheng Pan** [2]

[1] Nanjing Normal University, Nanjing, China
[2] NARI Group Corporation State Grid Electric Power Research Institute, Nanjing, China

[*] Corresponding author's e-mail: xuzhihua2006@126.com

Abstract. Due to long lines, insufficient redundancy, and low automation levels of distribution networks in remote areas in China, it is difficult to guarantee the reliability of power supply and power quality, and the vulnerability of power grids has increased after new energy photovoltaic access. These regional power grids have problems such as weak network topology, poor distribution line conditions, and poor communication conditions, which make it difficult for existing technologies to meet the needs of fault handling. This paper researches the operation fault of low-voltage power grids in remote areas, analyzes the characteristics of short circuit faults and small current grounding faults after the new energy photovoltaic is connected, and finds that its impact is closely related to the power supply type, access location, grounding mode and other factors, which provides new ideas and methods for solving low-voltage power grid fault in remote area.

1. Introduction

Due to the wide coverage, long power supply lines, insufficient redundancy, and low level of automation of the distribution network in remote areas of China, it is difficult to guarantee power supply reliability and power quality. With the rapid growth of new energy sources, these problems are further exacerbated, making the vulnerability of the power grid more obvious. The power grids in these areas usually have the following characteristics [1].

1) Chaotic line topology structure: the abnormal redundancy of the power grid is low, and some nodes lack standby power supply channels, leading to insufficient distributed photovoltaic access capability, and it is difficult to realize load transfer through network reconfiguration after the abnormality, which seriously affects power supply reliability.

2) Poor conditions of distribution lines: long transmission distance, large impedance, many branches and sectioning points of the line make it difficult to maintain the line, high fault rate, and difficult to repair, which increases the difficulty of fault location and rapid removal.

3) The communication condition is not ideal: the communication coverage is limited, the channel bandwidth is insufficient, the network delay is high and the connection is unstable, it is difficult to obtain complete, accurate, and timely key node information, which brings challenges to the system status monitoring and fault location. At the same time, large-scale access to distributed PV has brought more uncertainty to the power grid operation, which seriously threatens the security and stability of the system.

These characteristics make the fault handling of the distribution network face great challenges, and the existing distribution network state estimation, fault diagnosis, and fault recovery make it difficult to meet the demand after the new energy photovoltaic access.

In addition, the distribution networks in remote areas are mostly built in the early stage, and the equipment is severely aging, which makes power failures occur frequently. The monitoring points of the distribution network are widely distributed. The complex geographical environment, bad weather conditions, and natural disasters all bring great challenges to the maintenance of monitoring equipment

Journal of Physics: Conference Series **3031** (2025) 012015
doi:10.1088/1742-6596/3031/1/012015

and communication equipment. Interference, delay, and interruption are common in the process of data transmission. These problems not only affect the accuracy and real-time state estimation of the distribution network but also lead to deviation in the identification of weak links, which hinders the effective improvement and management of the distribution network and reduces the reliability and operating efficiency of the power supply [2].

Although some achievements have been made in the field of transmission line fault identification and detection in China, research on distribution line fault processing mainly focuses on technologies such as distance measurement and line selection, and systematic research on distribution line operation faults in remote areas is relatively scarce.

2. Short-circuit fault analysis after the integration of new energy photovoltaics

In the case of new energy photovoltaic access, it can be equivalent to the dual power supply system analysis model as shown in Figure 1. The system voltages on Side A and Side B are E_A and E_B respectively, and the equivalent impedances are Z_A and Z_B respectively, where E_A is the main grid power supply and E_B is the new energy photovoltaic.

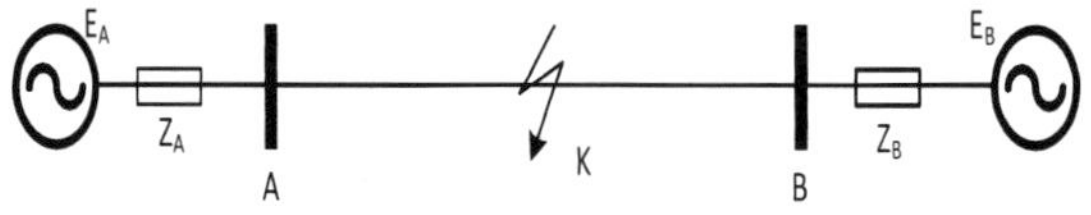

Figure 1. Equivalent dual terminal power supply system model with new energy photovoltaic.

In the case of a single-phase ground fault at Point F through the transition resistance, the measured voltage at Node K is taken as the reference basis, and the fault voltage phase angle Φ AK can be divided into two parts, one is the phase angle between the measured voltage and the measured current, and the other is the phase angle between the measured current and the fault voltage. The specific relationship is as Equation (1):

$$\arg\left(\frac{I_a}{U_k}\right) = \arg\left(\frac{I_a}{I_k * R_k}\right) = \arg\left(\frac{I_a}{I_k}\right) \tag{1}$$

Because the current at the fault point cannot be measured, it is difficult to calculate the phase angle between the measured current and the fault current. In case of a system fault, the phase of fault current is consistent with the phase of fault sequence current, and the phase of sequence current and fault sequence current can be approximately regarded as the same at the measurement installation, that is, the measured sequence current of Node K is in the same phase as the fault sequence current. It can be seen that the phase angle between the measured current and the fault voltage is approximately equal to the phase angle between the measured current and the sequence current of Node K [3], and the specific relationship is as f Equation (2):

$$\arg\left(\frac{I_a}{U_k}\right) = \arg\left(\frac{I_a}{C * I_k^2}\right) \approx \arg\left(\frac{U_a}{I_k^2}\right) \tag{2}$$

C is the negative sequence current distribution coefficient, and I^2_K is the negative sequence current at side K.

In general, the influence of DG on distribution network fault characteristics can be summarized as follows:

1) When a three-phase short circuit fault occurs upstream of the feeder where DG is located, DG will be decoupled from the power side circuit of the system, and at this time DG will not affect the short circuit current supplied by the system.

2) Under the condition of low fault degree, if the DG outputs active current, the voltage at the junction will be affected by the DG output power, and with the increase of DG output power, the voltage will rise first and then decline. When the fault is very serious and DG only outputs reactive current, the grid point voltage is affected by the rated power of DG. As the rated power of DG increases, it also increases first and then decreases.

Journal of Physics: Conference Series **3031** (2025) 012015 doi:10.1088/1742-6596/3031/1/012015

3) When the fault degree is low and the new energy photovoltaic (DG) outputs active current, the short circuit current upstream of the feeder where the DG is located will first decrease and then increase with the increase of DG access capacity; The short-circuit positive sequence current of the downstream feeder where DG is located and the adjacent feeder will increase first and then decrease with the increase of DG access capacity. When the fault is very serious and DG only outputs reactive current, the current flowing through all protective devices of the distribution network is only related to the rated power of DG, but not to the output power of DG [4].

To sum up, in the initial stage of the fault, the current characteristics show a state; As time goes on, the current shows another characteristic.

3. Analysis of small current grounding fault after new energy photovoltaic connection

3.1. Influence analysis of steady-state fault characteristics

When a small current grounding fault occurs in an active distribution network whose neutral point is grounded through an arc suppression coil, Phase A is taken as the fault phase, and a double outgoing line system is taken as an example. According to the superposition theorem, the original system can be decomposed into the superposition of three single power supply systems, as shown in Figure 2, where the arrow represents the flow direction of capacitive current. Specifically, Figure 2 (a) represents the original system, while Figure 2 (b), (c), and (d) respectively correspond to three single power supply systems. This decomposition method helps analyze the distribution and characteristics of fault current and provides a theoretical basis for fault location and protection strategy formulation.

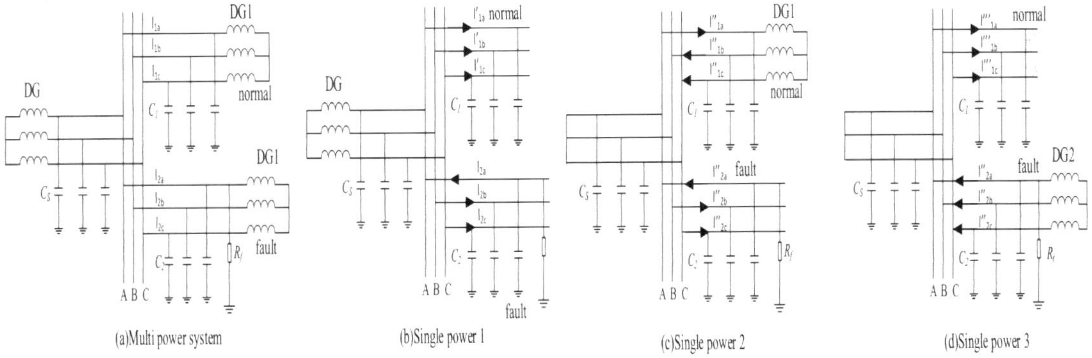

Figure 2. Active distribution network with ungrounded neutral and its decomposition model.

It can be seen that in case of fault, the magnitude of Phase B and Phase C current of the normal line will be affected, but the flow direction remains unchanged and still flows from the bus to the line; The magnitude of the A-phase current of the normal line will also change, and when the new energy photovoltaic in the normal line reverses its power supply to the grid, the direction of the A-phase current may reverse, that is, from the line to the bus.

The current of each phase at the head end of the fault line is:

$$\begin{cases} I_{na} = I'_{na} + I''_{na} + I'''_{na} \\ I_{nb} = I'_{nb} + I''_{nb} + I'''_{nb} \\ I_{nc} = I'_{nc} + I''_{nc} + I'''_{nc} \end{cases}$$

I_{na}, I_{nb}, and I_{nc} are the current of Phase A, Phase B, and Phase C at the head end of the original system line n respectively; I_{na}', I_{nb}', and I_{nc}' are the current of Phase A, Phase B, and Phase C at the head end of line n of single power supply system 1 respectively; I_{na}'', I_{nb}'', and I_{nc}'' refer to Phase A, Phase B, and Phase C currents at the head end of line n of the single power system 2 respectively; I_{na}''', I_{nb}''', and I_{nc}''' refer to the current of Phase A, Phase B, and Phase C at the head end of line n of the single power supply system 3, respectively, with n = 1.2.

The fault line is studied and the zero sequence current upstream and downstream of the fault point is analyzed.

$$\begin{cases} I_{2a.u/d} = I'_{2a.u/d} + I''_{2a.u/d} \\ I_{2b.u/d} = I'_{2b.u/d} - I''_{2b.u/d} \\ I_{2c.u/d} = I'_{2c.u/d} - I''_{2c.u/d} \end{cases}$$

u represents the upstream line and d represents the downstream line.

Under fault conditions, the current amplitude of the normal phase and the fault phase line upstream of the fault point will change, but the current direction will remain unchanged; The amplitude of the fault phase line current downstream of the fault point will also change. When the new energy photovoltaic at the end of the fault line supplies power to the grid in reverse, the direction of the line current will reverse.

When studying the zero-sequence network, when the new energy photovoltaic is connected to the grid, the high voltage side of the transformer at the grid connection adopts delta or Y-type wiring mode to ensure that the system neutral point remains the original grounding mode. From the perspective of safety and grid specification, the grounding mode of new energy PV in the small current grounding system should be ungrounded. In the active distribution network, when a single-phase ground fault occurs, the topology of the zero-sequence network is the same as when the new energy photovoltaic is not connected, so the connection of the new energy photovoltaic will not change the steady-state characteristics of the zero-sequence current [5].

3.2. Influence analysis of transient fault characteristics

When the rural distribution network with distributed generation fails, both the system power supply and the distributed generation DG will transmit fault current to the fault point. The type and location of DG will affect the transient characteristics of small current grounding faults in the distribution network. DG of distributed power supply is generally divided into rotary type and inverter type. The common rotating DG is a diesel generator, asynchronous motor, etc; Inverter DG mainly covers photovoltaic power generation and wind power generation.

Here we analyze the inverter type, and the fault current of the inverter DG is closely related to the inverter control strategy. In fault analysis of a distribution network with inverter DG, inverter DG is usually equivalent to a constant positive sequence current source [6]. Details are shown in Figure 3(a).

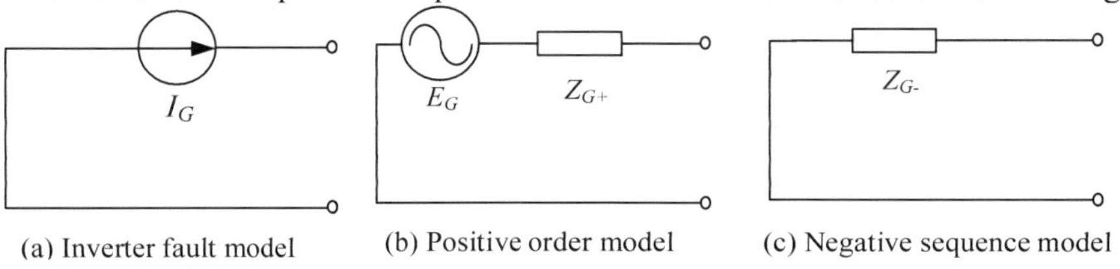

Figure 3. Rotating and Inverter DG Fault Model.

The transient characteristics of small current grounding faults in the distribution network, when DG is connected to different locations, are analyzed in depth. According to the difference in DG access location, it can be divided into the following three scenarios:

1) DG is located in the normal line: that is, the line connected by DG is not the fault line;

2) DG is located upstream of the fault point: that is, DG is located between the fault point and the bus;

3) DG is located downstream of the fault point: it means that DG is on the side away from the bus at the fault point.

The line mode equivalent circuit corresponding to these three scenarios is shown in Figure 3 (b)(c). When DG is upstream of the normal line or fault point, the equivalent impedance upstream of the fault point will change. When DG is located downstream of the fault point, it changes the equivalent impedance downstream of the fault point.

When DG is not connected, a single-phase grounding fault in the distribution network will generate a corresponding fault current. After the DG is connected, the actual fault current will increase slightly due to the action of $Z_{(1)DG}$ and $Z_{(2)DG}$, but the difference between the two is very small. It should be noted that after the DG is connected, the parameters of the zero-mode network will not be affected, and the impact on the zero-mode voltage can be ignored.

In addition, the influence of DG on fault transient current is similar to that on fault power frequency current. In other words, DG access may change the distribution of transient line mode current, but it will not affect the distribution of transient zero mode current.

4. Ground fault analysis technology based on multi frequency standard synchronous measurement

In the traditional rural radial distribution network, the steady-state and transient characteristics of various typical fault types, especially the small current grounding fault, are deeply analyzed.

The small current grounding fault of rural distribution networks has weak characteristic signals and complex and changeable fault scenarios, and the characteristics of different faults are quite different. The massive access of new energy photovoltaics has further changed the original power flow direction and fault characteristics, making the fault section location algorithm relying solely on steady or transient quantities face failure risk. With the rapid development of new-generation measurement technology, PMU, DTU, FTU, Transient recording type LTU, and other equipment provide rich waveform information for fault section location. By synchronously obtaining waveform information from multi-scale measurement nodes, and mining fault features from a global perspective, it can effectively remedy the shortcomings of traditional localization algorithms. Fault features can be extracted from multiple dimensions such as wide domain, frequency domain, and point domain. However, the more fault features extracted, the higher the requirements for the accuracy of measurement devices and the computing capacity of regional master stations. Moreover, different fault characteristics have different sensitivity to faults, so it is necessary to screen fault characteristics and extract better fault characteristics to characterize the change law of electrical quantities at measurement points when faults occur in different sections.

The core idea of the small current grounding fault location method based on multi-point synchronous waveform and matrix calculation is to extract the synchronous waveform characteristics of multiple measurement points from the perspective of multiple domains. With the help of random matrix theory, the analysis object is converted from the fault feature matrix to the random matrix eigenvalue, and the historical samples are used to determine the cluster center point of the fault section in the eigenvalue space. When an online fault location is carried out, the fault section can be determined according to the matching result of the eigenvalue space distance. The specific process of this method is as follows, refer to Figure 4:

(1) Fault feature extraction and filtering:

1) Starting from the historical fault recording data, the fault characteristics are calculated respectively in the time domain, frequency domain, and time-frequency domain, and the fault characteristics $Z_m = [z_{m1}, z_{m2}, \ldots, z_{mk}]$.

2) The method of integrating the Relief F algorithm and RF algorithm is used to comprehensively weigh the calculated fault characteristics.

3) Set the weight dead zone wdb _ f, filter out the k-dimension fault features with the comprehensive weight greater than wdb _ f, and retain them.

(2) Determine the location of the cluster center point:

1) For each historical fault sample, the topology matrix is constructed one by one.

2) The random matrix algorithm is used to extract the eigenvalues from the topological matrix, and the first three-dimensional eigenvalues are selected as the location coordinates of each fault sample in the L-space space.

Journal of Physics: Conference Series **3031** (2025) 012015 doi:10.1088/1742-6596/3031/1/012015

3) Search the central point of the cluster, and then determine the corresponding central point position of each section in the L-space space [7].

(3) Real-time fault feature processing:

1) When the system detects a single-phase ground fault, k fault characteristics of transient zero mode current in the time domain, frequency domain, and time-frequency domain are calculated locally at each measuring point.

2) After receiving the data, the regional master station normalizes the fault characteristics of each measurement point and generates a topology matrix.

3) The topological matrix is transformed into a random matrix, and then the eigenvalues of the random matrix are obtained.

(4) Fault section location:

1) The first four-dimensional eigenvalue of the random matrix is extracted to determine its position in the L-space space.

2) This position is similar to the center point of each fault section in the L-space space, and the section closest to the center point is the fault section.

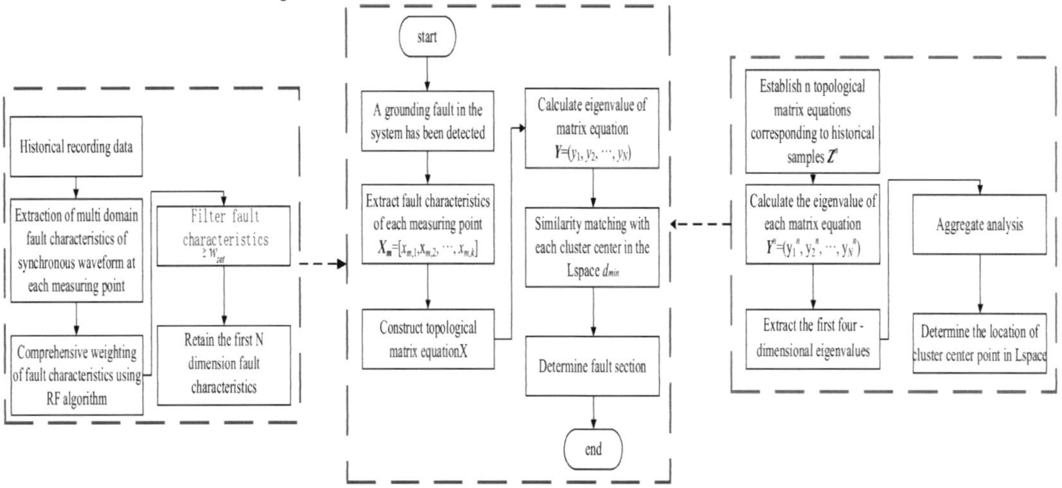

Figure 4. Fault handling of multi-frequency scale synchronous measurement and matrix calculation.

5. Case analysis

In the simulation environment, the low-voltage model shown is built in Figure 5. During modeling, various factors such as fault type, fault initial phase angle, transition resistance, and fault phase are comprehensively considered to generate corresponding fault information.

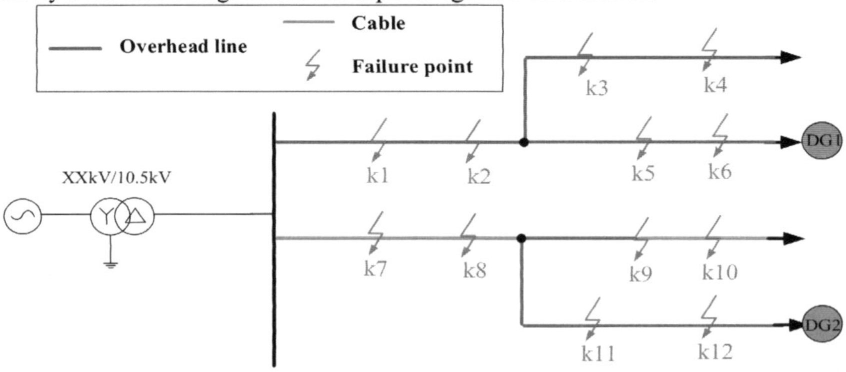

Figure 5. Low voltage topology and fault location diagram.

The final test results are summarized in Table 1, from which it can be seen that the recognition accuracy of the algorithm for single-phase ground fault reaches 96%; In terms of overall performance, it is at a high level. Simulation results show that the algorithm is less affected by fault phase, fault location, transition resistance and fault initial phase angle.

Table 1. Test result statistics.

Fault Type	Fault phase	Number of test cases	Accuracy
Single phase grounding	AG	400	385
	BG	400	383
	CG	400	384
total		1200	1152

6. Conclusion

This paper studies the operation fault of distribution network in remote areas, analyzes the characteristics of short circuit fault and small current grounding fault after the new energy photovoltaic is connected, and puts forward the grounding fault processing technology based on multi frequency synchronous measurement. The grounding network fault location technology based on multi frequency synchronous measurement, uses the waveform information provided by the new generation of measurement technology, synchronously obtains the waveform from the multi-scale measurement nodes, and excavates the fault characteristics, which can effectively remedy the defects of the traditional location algorithm, realize the rapid location of fault lines, improve the accuracy and real-time of fault processing, and provide new ideas and methods for solving the problem of fault location of distribution network in remote areas. It has important practical value.

References

[1] Zhao et al. Research and application of low-voltage distribution network comprehensive treatment equipment under large-scale distributed photovoltaic access [J]. Rural electrification, 2024, (12):1-5.
[2] Zheng et al. Application of new relay protection connected to distributed power distribution system [J]. Electrical Application, 2015, 34 (S2): 287-290.
[3] Chen et al. A traveling wave natural frequency-based single-ended fault location method adaptive to system impedance[J]. Power System Technology, 2013, 37(6):1739-1745.
[4] Liu et al. Research on Regional Autonomy Collaborative Control Technology of Pan microgrid Based on Multi-Agent Technology [J]. Power Grid and Clean Energy, 2022, 38 (12): 61-71.
[5] Liu et al. Distribution network fault recovery method including distributed generation access [J]. Electric Measurement and Instrumentation, 2020, 57 (18): 50-56.
[6] Xu et al. Design and application of distribution network multi-function terminal based on synchronized phasor measurement information [J]. Electrical Application, 2020, 39 (01): 88-93.
[7] Wang et al. PMU optimal configuration of active distribution network with observable weak nodes [J/OL]. Journal of Electrical Engineering, 1-12 [2025-02-08].

Journal of Physics: Conference Series **3031** (2025) 012016 doi:10.1088/1742-6596/3031/1/012016

Assessment of dynamic reactive power reserve requirements in power systems based on transient voltage severity index

Chong Li [1], Jinquan Zhao [1*], Jie Dang [2], Ting Cui [3], Daojun Chen [3]

[1] College of Electrical Engineering, Shanghai University of Electric Power, Shanghai, China
[2] Central China Branch of State Grid, Wuhan, China
[3] State Grid Hunan Electric Power Company Limited Research Institute, Changsha, China

[*] Corresponding author's e-mail: zhaojinquan@shiep.edu.cn

Abstract. With the increasing integration of renewable energy sources and converter-interfaced resources, the dynamic reactive power reserve (DRPR) for ensuring transient voltage stability in modern power systems has diminished, necessitating a systematic approach to quantify its minimum requirement. This paper proposes a novel DRPR assessment method based on the transient voltage severity index (TVSI). First, the power grid is partitioned into coherent regions by using an agglomerative hierarchical clustering algorithm, grouping nodes with similar transient voltage characteristics derived from TVSI. Second, dynamic adjustments to DRPR are iteratively applied within each partition to bring the TVSI of the most voltage-weak node close to the stability threshold. Finally, the critical-state DRPR of each partition, calculated by aggregating sensitivity-weighted reserves from reactive power sources, is defined as the minimum requirement. Validation of the CSEE-VS system demonstrates the effectiveness of the proposed method in identifying DRPR requirements.

1. Introduction

New power systems are undergoing transformative changes driven by the integration of renewable energy sources, distributed generation, and advanced ultra-high voltage direct current (UHVDC) transmission technologies [1]. These developments have introduced significant challenges to transient voltage stability due to the declining availability of dynamic reactive power reserves (DRPR) [2]. DRPR plays a pivotal role in mitigating voltage instability during contingencies, such as short circuits or sudden load changes, by providing rapid reactive power support [3]. However, the growing reliance on converter-interfaced resources (e.g., wind and solar farms) has diminished traditional synchronous generator-based DRPR capacity, creating a critical need to reassess the minimum DRPR requirements for ensuring transient voltage stability.

The definition of DRPR lacks standardization. Traditional DRPR definitions typically quantify reserves as the difference between the maximum reactive power output of devices (e.g., generators, SVGs) and their steady-state operating values [4, 5]. However, there exists a possibility that the reactive power source may not reach its maximum reactive power output before the occurrence of a voltage collapse in the system. Consequently, this definitional approach may lead to overly optimistic assessment results. Therefore, dynamic effective reactive power reserves are defined [6-9]. The upper limit of the reactive power output for a reactive power source is defined as the actual reactive power output during the most severe fault conditions [6, 7], or the reactive power output at the point of voltage collapse [8, 9]. The dynamic reactive power reserve definitions described above are for static voltage stability, relying on static analytical methods, and lacking consideration of changes in reactive power output from reactive power sources during transient processes. Recent studies have proposed transient-oriented DRPR definitions to address the limitations of static approaches. These definitions emphasize the effectiveness of reserves during transient events [10, 11]. In [10], DRPR is defined as the sum of the reactive power incremented by the reactive source at each moment after a fault, multiplied by a correction factor for valuation. However, this definition does not specify the minimum requirement for DRPR. In [11], the increment in the maximum reactive power output of the generator corresponding to the second dip in the converter bus voltage is defined as the effective DRPR, with the minimum requirement specified. However, this approach only addresses the assessment needs related to the failure of the DC continuous current commutation and does not adequately account for transient voltage stability.

To address these limitations, this paper proposes a novel method for assessing DRPR requirements based on the transient voltage severity index (TVSI). First, the power grid is partitioned into coherent regions by using an agglomerative hierarchical clustering algorithm that groups nodes with similar transient voltage characteristics, ensuring localized reserve allocation. Second, dynamic adjustments to DRPR are implemented within each partition, targeting the node exhibiting the most severe voltage dip. By iteratively tuning the reserve until the TVSI approaches a predefined threshold, the method identifies the critical DRPR level required to stabilize the system. Finally, the reactive power reserves within the partition are aggregated by using a weighted sum based on reactive power/reactive power sensitivity to obtain the partition DRPR. The partition DRPR under the critical state is defined as the minimum requirement for that partition. The effectiveness of the proposed method is validated in the CSEE-VS system.

2. Power system partitioning method based on transient voltage severity index

Reactive power in a power system cannot be transmitted over long distances and is inherently regionally balanced. Even when the system possesses a substantial DRPR, voltage instability may arise due to its uneven distribution. Therefore, the assessment of the DRPR requirements should be conducted on a partitioned basis. This paper establishes node-level transient voltage characteristics derived from the TVSI. An electrical distance metric is formulated based on the similarity of these characteristics, followed by the application of an agglomerative hierarchical clustering algorithm to delineate power grid partitions.

2.1. Transient voltage severity index

TVSI is an index that describes the degree of node voltage dip and can also reflect the transient voltage characteristics of the node. In practical engineering, transient voltage that dips below a specified threshold U_{cr} and persists for a certain duration T_{cr} is classified as transient voltage instability. $[U_{cr}, T_{cr}]$ is defined as a two-element notation. When multiple-two-element notations $\{[U_{cr,1}, T_{cr,1}], \cdots, [U_{cr,k}, T_{cr,k}], \cdots, [U_{cr,n}, T_{cr,n}]\}$ exists, as long as one of them is not satisfied, transient voltage instability can be determined. TVSI based on multiple-two-element notations is defined as follows [12]:

$$\Lambda_{i,j,k} = \frac{S_k}{S_{cr,k}} = \frac{\int_{t_k}^{\tau_k} [U_N - U_i(t)]dt}{(U_N - U_{cr,k})T_{cr,k}} \quad k = 1, 2, \cdots, n \tag{1}$$

where $\Lambda_{i,j,k}$ is the TVSI of node i under fault j corresponding to two-element notation $[U_{cr,k}, T_{cr,k}]$; S_k is the area of the corresponding sub-region of $[U_{cr,k}, T_{cr,k}]$; $S_{cr,k}$ is the critical area of $[U_{cr,k}, T_{cr,k}]$; $U_i(t)$ is the voltage at node i at the moment of t; U_N is the nominal voltage magnitude; t_k is the time when $U_i(t) < U_{cr,k}$; τ_k is the time when $U_i(t) > U_{cr,k}$.

TVSI quantifies the severity of transient voltage dips by comparing S_k and $S_{cr,k}$. Notably, while the index is defined by multiple-two-element notations, the transient voltage dip severity at a node is not evaluated via a composite metric. Instead, it is characterized by a set of values equal to the number

Journal of Physics: Conference Series **3031** (2025) 012016 doi:10.1088/1742-6596/3031/1/012016

of two-element notations. If $\Lambda_{i,j,k} < 1$, node i satisfies the requirement of $[U_{cr,k}, T_{cr,k}]$ under fault j. Node i is deemed transient voltage stable under fault j only if $\Lambda_{i,j,k} < 1, k = 1, 2, \cdots, n$. Conversely, if any $\Lambda_{i,j,k} > 1$, node i is identified as a transient voltage-weak node under fault j, indicating transient voltage instability.

2.2. *Similarity of transient voltage characteristics*

TVSI serves as a characteristic of node transient voltage while evaluating node transient voltage stability based on voltage curves. The voltage curves vary under different fault conditions. To provide a more accurate description of node transient voltage characteristics, node transient voltage characteristic vectors are constructed following a fault scanning of the grid:

$$\mathbf{A}_i = \left\{ \Lambda_{i,j,k} \mid j = 1, 2, \cdots, N_J; k = 1, 2, \cdots, n \right\} \tag{2}$$

where $\mathbf{A}_i$ is the transient voltage characteristic vector of node i and N_J is the number of faults in the predefined fault set.

Cosine similarity is a metric for measuring the similarity between two vectors, determined by calculating the cosine of the angle between them, which allows for the assessment of vector similarity independent of their magnitudes. In this paper, cosine similarity is employed to quantify the similarity of transient voltage characteristics between nodes, serving as a measure of transient electrical distance [13]:

$$d_{i,j} = \frac{|\mathbf{A}_i||\mathbf{A}_j|}{\mathbf{A}_i \mathbf{A}_j^T} \tag{3}$$

After obtaining the electrical distances between nodes, the grid is partitioned into N_{area} regions based on a cohesive hierarchical clustering algorithm [14].

3. Assessment of dynamic reactive power reserve requirements

3.1. *Definition of dynamic reactive power reserve*

To reflect the transient effectiveness of DRPR, the definition of DRPR in this paper is based on the actual reactive power output of the reactive power source after the fault, which is expressed as follows:

$$Q_{rs,i,j} = \frac{1}{t_{m,i,j} - t_{c,j}} \int_{t_{c,j}}^{t_{m,i,j}} [Q_{i,j}(t) - Q_{i0}] \mathrm{d}t \tag{4}$$

where $Q_{rs,i,j}$ is the DRPR of the reactive power source i under fault j; $t_{c,j}$ is the fault-clearing moment; $t_{m,i,j}$ is the moment when the reactive power output of the reactive power source i rises to the maximal reactive power output; $Q_{i,j}(t)$ is the reactive power output of the reactive power source i at the moment of t; Q_{i0} is the initial reactive power output of the reactive power source i.

The physical significance of dynamic reactive power reserve, as defined in this paper, refers to the actual average incremental reactive power generated by a reactive power source during the interval between fault clearance and the attainment of its maximum reactive power output under a specified fault condition.

This paper assesses the DRPR of the system at the partition level. Directly summing the DRPR of individual reactive power sources within a partition fails to account for their contributions to stabilizing voltage-weak nodes. To address this limitation, the proposed method calculates the partition-level DRPR by weighting the DRPR of each source with its reactive power-to-reactive power sensitivity relative to voltage-weak nodes, followed by summation. The mathematical formulation is expressed as:

$$Q_{RS,z,j} = \sum_{j=1}^{N_{g,k}} \frac{\Delta Q_{i_{w,j,z}}}{\Delta Q_i} Q_{rs,i,j} \tag{5}$$

where $Q_{RS,z,j}$ is the DRPR of partition z under fault j; $N_{g,k}$ is the number of dynamic reactive power sources in partition z; $\Delta Q_{i_{w,j}} / \Delta Q_i$ is the reactive power-to-reactive power sensitivity of the reactive demand of node $i_{w,j,z}$ to the reactive power output of the reactive power source node j, which is obtained by calculating based on the numerical sensitivity method; $i_{w,j,z}$ is the most voltage-weak node in partition z under fault j, which is maximized by screening TVSI:

$$\Lambda_{i_{w,j,z}} = \max\{\Lambda_{i,j,k} \mid i = 1,2,\cdots,N_z, k = 1,2,\cdots,n\} \tag{6}$$

where N_z is the number of nodes in partition z.

3.2. *Dynamic reactive power reserve adjustment measures*

This paper defines the DRPR under the critical state of transient voltage stability as the minimum requirement. The critical state is determined by TVSI. Specifically, a partition is identified as reaching the critical state when its maximum TVSI ($\Lambda_{i_{w,j,z}}$) approaches 1. To achieve this state, iterative adjustments to the DRPR are applied until $\Lambda_{i_{w,j,z}}$ approaches 1.

The specific dynamic reactive power reserve regulation measures and their prerequisites are outlined as follows: when $\Lambda_{i_{w,j,z}} > 1$, the partition z is deemed to lack sufficient dynamic reactive power reserve, necessitating an increase. A fixed-capacity Static Var Generator (SVG) should be installed at the node with the weakest voltage, followed by a recalculation. Conversely, when $\Lambda_{i_{w,j,z}} < 1$, the partition z is considered to have an adequate dynamic reactive power reserve, requiring a reduction. It is recommended to decrease the reactive power output capacity of the reactive power source that exhibits the lowest sensitivity to reactive power changes at the weakest voltage node, followed by a recalculation.

By calculating the DRPR for each partition following the application of the adjustment, the minimum requirement for DRPR in each partition can be determined. Subsequently, by comparing this with the initial DRPR of each partition, the dynamic reactive power reserve level of the system can be assessed.

4. Example analysis

4.1. *Grid partition*

In this paper, the CSEE-VS system [15] is utilized to validate the effectiveness of the proposed evaluation method, with the system structure illustrated in Figure 1. The multiple-two-element notation is set up to be {[0.75 p.u., 1 s], [0.9 p.u., 10 s]}. The curves depicting the variation of merging distances with the number of partitions at each step are presented in Figure 2, while the partitioning results are annotated in Figure 1.

Three representative nodes were selected from each partition to validate the homogeneity of transient voltage characteristics. Their voltage curves under the B03-B05 N-2 fault are illustrated in Figure 3, while the values of TVSI are summarized in Table 1. The results demonstrate that nodes within the same partition exhibit similar voltage trajectory morphologies during transient events, characterized by synchronized dip magnitudes and recovery rates. This coherence is further corroborated by the similarity in TVSI values across intra-partition nodes. These findings conclusively validate the effectiveness of the agglomerative hierarchical clustering algorithm in grouping nodes with homogeneous dynamic behaviors, thereby establishing a robust foundation for dynamic reactive power reserve assessment.

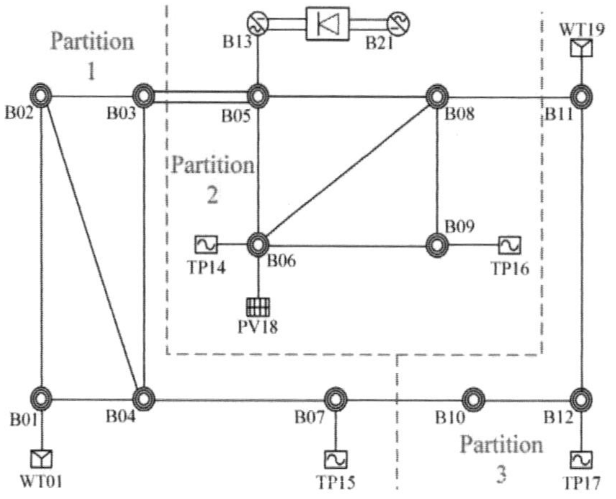

Figure 1. CSEE-VS system with three partitions.

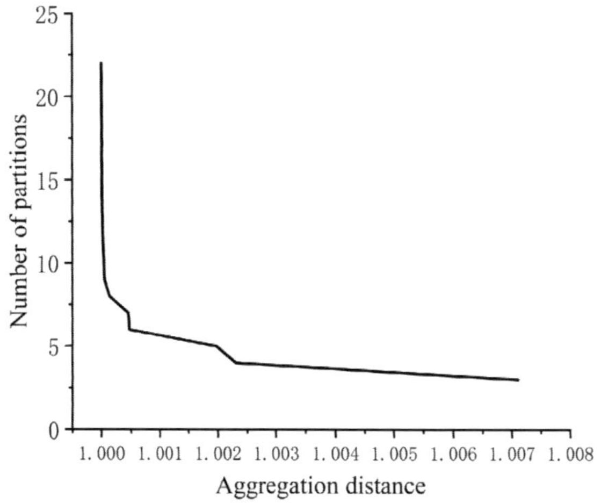

Figure 2. Clustering curves of CSEE-VS system.

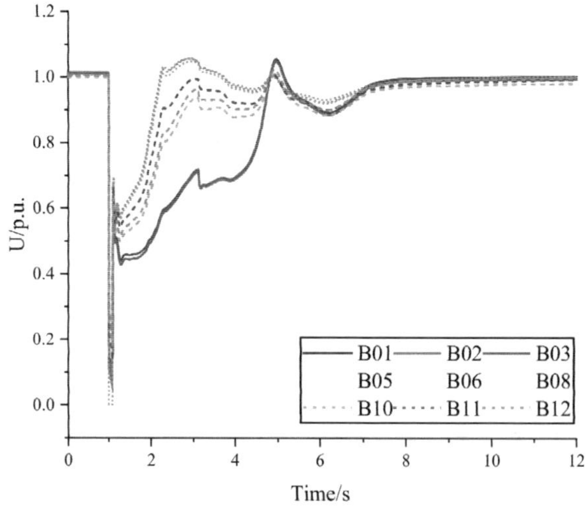

Figure 3. Voltage curves of representative nodes in each partition.

Table 1. TVSI for representative nodes in each partition.

Partition	1			2			3		
Node	B01	B02	B03	B05	B06	B08	B10	B11	B12
TVSI [0.75 p.u., 1 s]	5.37	5.56	5.50	0.87	0.86	0.89	2.16	1.95	2.02
TVSI [0.9 p.u., 10 s]	0.84	0.89	0.88	0.30	0.29	0.31	0.63	0.55	0.59

4.2. Assessment of dynamic reactive power reserve requirements

The power system should have sufficient dynamic reactive power reserves to ensure stable operation under all anticipated faults. Therefore, the DRPR requirement of the system under the most severe anticipated faults is the minimum requirement of the system to ensure stable operation. In this paper, the B03-B05 N-2 fault is identified as the most severe anticipated fault, and the minimum DRPR requirement of each partition is assessed under this fault.

To better validate the effectiveness of the proposed method, the DRPR requirements of the CSEE-VS system are assessed under multiple scenarios. Considering that induction motor loads are the dominant reactive loads during transient processes, this study examines CSEE-VS systems with varying induction motor load shares. The load shares considered in different scenarios are 30%, 35%, and 40%, respectively.

Under B03-B05 N-2 fault, TVSI for the most voltage-weak node in each partition with different induction motor load shares are shown below:

Table 2. TVSI for the most voltage-weak node with different induction motor load shares.

Partition	$i_{w,j,z}$	$\Lambda_{i_{w,j,z}}$ with different induction motor load shares		
		30%	35%	40%
1	B02	5.56	12.83	14.22
2	B08	0.89	2.48	3.95
3	B10	2.16	5.39	7.60

According to the calculation results of the transient voltage severity index in Table 2, it is evident that transient voltage stability deteriorates as the induction motor load share increases. Partitions 1 and 3 consistently lack dynamic reactive power reserves across all scenarios. Partition 2 has sufficient dynamic reactive power reserves only when the induction motor load share is 30%, but as the load share increases, the reserves become insufficient. The results of DRPR calculation in the initial state are shown in Table 3

Table 3. The results of DRPR calculation in the initial state.

Partition	Reactive power source	Units type	DRPR/Mvar	
			Reactive power source	Partition
1	WT01-01	Wind farm	87.94	464.92
	WT01-02		108.85	
	WT01-03		87.94	
	TP15-01	Synchronous generators	580.95	
2	SC13-01	Synchronous compensator	543.47	1610.45
	SC13-02		543.47	
	TP14-01	Synchronous generators	285.56	
	TP16-01		370.32	
	PV18-01	Photovoltaic power station	91.81	
	PV18-02		91.81	
	PV18-03		91.81	
3	TP17-01	Synchronous generators	444.00	239.99
	WT19-01	Wind farm	78.59	
	WT19-02		96.93	
	WT19-03		78.59	

Journal of Physics: Conference Series **3031** (2025) 012016 doi:10.1088/1742-6596/3031/1/012016

Using partition 1 as a case study, this section demonstrates the DRPR adjustment process. SVGs with a rated capacity of 100 Mvar are incrementally added at the most voltage-weak node until TVSI approaches 1. The iterative calculation process with different induction motor load shares is detailed in Table 4.

Table 4. Iterative calculation process with different induction motor load shares.

Number of iterations	$\Lambda_{i_{w,j,z}}$ with different induction motor load shares		
	30%	35%	40%
0	5.55	12.83	14.22
1	3.44	8.45	11.20
2	2.76	5.88	10.42
3	2.26	4.98	8.64
4	1.95	3.84	5.67
5	1.60	3.57	3.92
6	1.24	2.34	2.98
7	0.99	1.43	2.43
8		0.97	1.97
9			1.31
10			0.99

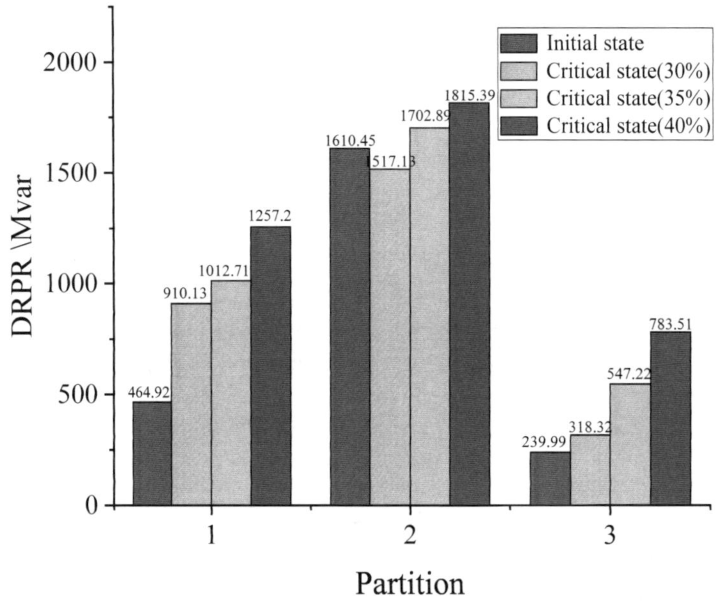

Figure 4. Comparison chart of DRPR and its requirement in each partition.

Upon determining the critical state, the minimum DRPR requirements can be calculated, with the results presented in Figure 4. These requirements are then compared to the DRPR in the initial state, as illustrated in Figure 4. DRPR assessment can be effectively conducted. In the scenario where the induction motor load share is 30%, partitions 1 and 3 exhibit a deficiency in DRPR, with partition 1 exhibiting a greater degree of shortage. Conversely, partition 2 demonstrates sufficient dynamic reactive power reserves. However, as the proportion of induction motor load increases, the deficiencies in DRPR in partitions 1 and 3 become exacerbated, while partition 2 transitions from a state of sufficient DRPR to one of insufficiency.

The increasing penetration of induction motor loads elevates the system's reactive power demand during transient events. This phenomenon is corroborated by the results of the DRPR demand assessment in this paper, which further validates the effectiveness of the proposed method.

5. Conclusion

This paper addresses the critical challenge of assessing dynamic reactive power reserve (DRPR) requirements for transient voltage stability in power systems. The proposed methodology integrates transient voltage severity analysis, grid partitioning, and sensitivity-based reserve aggregation to deliver two key contributions:

(1) By leveraging TVSI and cosine similarity metrics, the power grid is partitioned into regions with homogeneous dynamic behaviors. This spatial decomposition enables localized DRPR assessment, addressing the inherent limitations of monolithic grid models.

(2) The minimum DRPR requirement is rigorously defined as the reserve required to maintain the TVSI of the most voltage-weak node in each partition at the stability threshold. This threshold-driven approach ensures that the derived reserves are both necessary and sufficient to ensure voltage stability.

Validation of the CSEE-VS system demonstrated the effectiveness of the proposed method. This research provides a scalable and data-driven foundation for DRPR planning in future power systems. Future research will focus on integrating the proposed method with artificial intelligence techniques to enhance its efficiency.

Acknowledgments

This work was financially supported by the Key Science and Technology Project of the State Grid Corporation of China (No. 5100-202323008A-1-1-ZN).

References

[1] Z. Zhang, C. Kang. "Challenges and prospects for constructing the new-type power system towards a carbon neutrality future," Proceedings of the CSEE, vol. 42, no. 8, pp. 2806-2818, 2022.
[2] M. Habib, A. Gram, A. Harrag, et al. "Optimized management of reactive power reserves of transmission grid-connected photovoltaic plants driven by an IoT solution," International Journal of Electrical Power & Energy Systems, vol. 143: 108455, 2022.
[3] N. Rani, T. Malakar. "Assessment of effective reactive power reserve in power system networks under uncertainty applying coronavirus herd immunity optimizer (CHIO) for operation simulation," Electric Power Systems Research, vol. 220: 109267, 2023.
[4] E. Davoodi, F. Capitanescu, L. Wehenkel. "A Methodology to Evaluate Reactive Power Reserves Scarcity During the Energy Transition," IEEE Transactions on Power Systems, vol. 38, no. 5, pp. 4355-4368, 2023.
[5] S. Rizvi. "A novel distance-based system reactive power index for system-level reactive power reserve assessment," IET Smart Grid, vol. 6, no. 3, pp. 259-270, 2023.
[6] Y. Qiu, H. Wu, Y Song, et al. "Optimization method for generator reactive power reserve considering static shunt reactive power compensation," Automation of Electric Power Systems, vol. 39, no. 18, pp. 31-36, 2015.
[7] G. Chen, W. Wu, Z. Dai, et al. "Multi-objective Optimization of Regional Reactive Power Reserve in Hybrid System with Wind, Photovoltaic and Energy Storage Considering Fault Scenario Set," Automation of Electric Power Systems, vol. 46, no. 17, pp. 194-204, 2022.
[8] J. Zhao, L. Ju, W. Luo, et al. "Reactive voltage control model and method considering partitioned dynamic reactive power reserve," Electric Power Automation Equipment, vol. 35, no. 5, pp. 100-105, 2015.
[9] S. Fang, H. Zhong, G. Xu, et al. "An optimization approach to generator reactive power reserve based on the closest critical point of voltage stability limit surface," Automation of Electric Power Systems, vol. 40, no. 3, pp. 27-32, 2016.
[10] Q. Sun, H. Cheng, J. Zhang, et al. "A dynamic reactive power reserve optimization method to enhance transient voltage security," Proceedings of the CSEE, vol. 35, no. 11, pp. 2718-2725, 2015.
[11] Y. Jiang, Y. Bao, J. Zhang, et al. "Dynamic reactive power reserve evaluation for DC continuous commutation failure," Power System Protection and Control, vol. 49, no. 19, pp. 173-180, 2021.
[12] Z. Zhao, J. Zhao, R. Shan, et al. "Transient Voltage Security Preventive Control Method for HVDC Receiving-End Power Grid," Southern Power System Technology, vol. 17, no. 3, pp. 1-9, 2023.
[13] L. Guan, L. Wu, Y. Zhuo, et al. "Large Power System Partitioning Method for Transient Voltage Control," Power System Technology, vol. 42, no. 11, pp. 3735-3759, 2018.
[14] J. Zhao, F. Liu, F. Liu, Y. Deng, et al. "Network Partitioning for Reactive Power/Voltage Control Based on a Mapping Division Algorithm," Automation of Electric Power Systems, vol. 34, no. 7, pp. 36-39, 2010.
[15] B. Zhao, S. Xu, T. Lan, et al. "Benchmark for AC-DC hybrid system with high penetration of renewables (3): voltage stability benchmark CSEE-VS," Proceedings of the CSEE, vol. 44, no. 21, pp. 8353-8364, 2024.

Journal of Physics: Conference Series **3031** (2025) 012017 doi:10.1088/1742-6596/3031/1/012017

Analysis of the future optimization path for low-carbon transition through energyplan software: the case of guangxi province in china

Liulin Yang [1], Shufan Tan [1], Yao Li [1], Tianlu Luo [2*]

[1]College of Electrical Engineering, Guangxi University, Nanning 530004, China
[2]Guangxi Power Grid Co., Ltd., Nanning 530023, China

[*]Correspondence: 1249184567@qq.com

Abstract. This paper provides optimization pathways for the future energy system in Guangxi Province, China. It aims to address the inefficient energy production and consumption structure. We established a 2050 quantitative transition model based on EnergyPLAN software. Due to the limitations of a quantitative model based on historical data, it cannot encompass all possible future scenarios, and there may be subjective biases in data processing. Thus, we coupled the model with the improved Gray Wolf Optimization algorithm to establish a long-term dynamic optimization scenario. This approach enables multiple iterations of interaction between modeling and optimization, achieving the goal of minimizing the total system cost and sustainable development. The results indicate that clean energy will satisfy approximately 84.2% of the energy demand in Guangxi's future energy structure. Compared to the quantitative transition scenario, the optimized scenario can save 18.5% of the annual total system cost while maintaining similar levels of carbon dioxide emissions. This provides new ideas for energy planning in Guangxi.

1. Introduction

The current greenhouse effect has resulted in global climate anomalies and rising sea levels, posing a significant threat to human survival and sustainable social development. The ongoing climate change crisis presents the greatest challenge of reducing greenhouse gas emissions. In China, carbon emissions from electricity production account for 40% of total emissions [1]. Integrated energy planning is a multi-objective optimization problem that encompasses economic, technological, and environmental aspects. Planning energy systems with a high share of renewable energy sources is becoming increasingly important. The Guangxi regional government has developed a series of energy transition plans aimed at advancing the regional energy system over the next decade. A key aspect of achieving this energy transition is upgrading and optimizing the existing energy supply and consumption structure. Energy system modeling has emerged as a primary solution to this challenge [2, 3].

2. Methods and materials

2.1 System optimization principles

EnergyPLAN serves as an energy system analysis tool for studying and designing solutions for future high-penetration sustainable energy planning [4-6]. It is a deterministic input/output tool, encompassing several critical parameters related to the energy system, including power, heating, cooling, industry, transportation, and desalination. Despite the powerful capabilities of EnergyPLAN for energy system modeling, its deterministic nature limits its ability to dynamically optimize different technologies and energy sources within the system. Therefore, coupling EnergyPLAN with optimization algorithms can effectively address these issues, facilitating a comprehensive and sustainable optimization of energy systems [7-9]. Grey Wolf Optimization (GWO) has demonstrated competitive results compared to well-known metaheuristic algorithms. This heuristic was developed by first translating collaborative behavior into an algorithm and subsequently into code, leading to rapid solutions for many engineering problems [10]. The GWO algorithm enables researchers to comprehensively assess the potential future impacts of proposed solutions while accounting for uncertainties in a given area. The algorithm exhibits strong convergence properties, allowing it to quickly and reliably converge to practical solutions after appropriate parameter adjustments [11, 12].

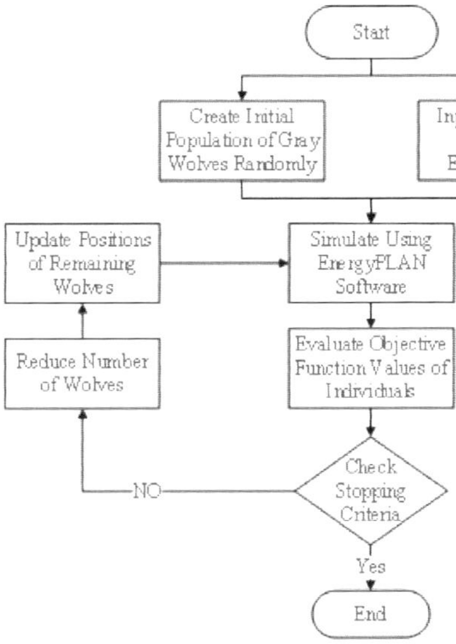

Figure 1. Operation flow chart of GWO coupled with EnergyPLAN.

In this study, these elements will be selected for modeling in EnergyPLAN. The coupled optimization process follows these steps, and the algorithm flow is shown in Figure 1:

- Establishment of the optimization objective in MATLAB: A random set of initial agent data is generated, which forms the initial "wolf pack" in the GWO algorithm.
- Evaluation using EnergyPLAN: The initial proxy data parameters, along with other necessary data parameters for modeling, are submitted to EnergyPLAN for evaluation, resulting in a complete energy configuration scenario.
- Model performance evaluation: The target performance of the model is assessed, and the results are fed back to the MATLAB platform. The optimization objective is achieved through the GWO algorithm, aiming to minimize the total annual cost of the energy system while meeting constraints for CO_2 emissions and import/export balance.
- Ranking and updating agents: The agents are ranked to identify the best solution. Based on these results, each agent's position is updated to form a new "wolf pack". The adjusted agents are then sent back to EnergyPLAN for further evaluation. This iterative process continues until predefined stopping criteria are met or the optimization is manually canceled.

2.2 Decision variables

Due to the inclusion of multiple sectors in the EnergyPLAN software, all these inputs can theoretically be viewed as potential decision variables in the coupling optimization process. Given the core position of the power sector within the overall energy structure, this paper focuses on optimizing and analyzing the power sector. We select the installed capacities of onshore wind power, offshore wind power, photovoltaic power, nuclear power, and pumped storage as the decision variables for optimization. The lower limits of these decision variables are based on the current actual installed capacities, while the upper limits depend on the potential maximum technical capacity and related policies. The range of these decision variables is illustrated in Figure 2.

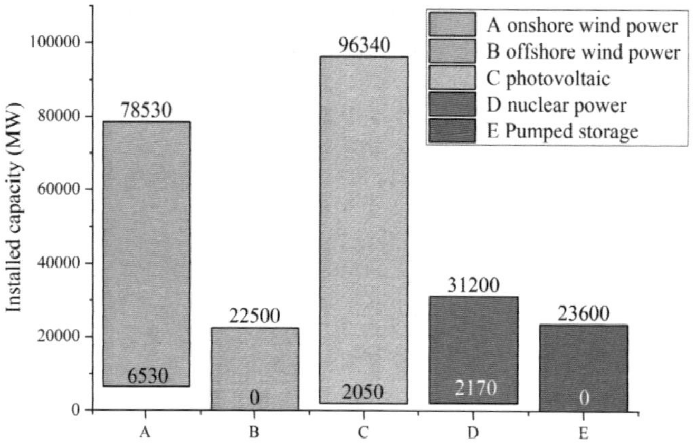

Figure 2. Range of decision variable installed capacity values.

2.3 Modeling parameters

In calculating the carbon emissions of the energy system, determining the carbon emission factors is the first step. This study adopts the carbon emission factors used in the national models of Denmark and China from the EnergyPLAN website, with specific values detailed in Table 1.

Table 1. Major fossil fuel CO_2 emission factors.

Types of fossil fuels	Coal	Oil	Natural gas	Liquefied petroleum gas	Waste
Carbon emission factors (kg/GJ)	95.0	74.0	56.7	59.6	32.5

Due to inconsistencies arising in statistical data due to the use of various energy units, such as tons of standard coal (tce), kilowatt-hours (kWh), or joules (J). Therefore, it is necessary to convert all energy input amounts to a unified unit in terawatt-hours.

Table 2. Reference coefficients of standard coal converted from various energy sources.

Energy types	Standard coal equivalent coefficient	Units
Raw coal	0.7143	Kilograms of standard coal/ Kilogram
Washed coal	0.9000	
Coke	0.9714	
Crude oil	1.4286	
Fuel oil	1.4286	
Gasoline	1.4714	
Kerosene	1.4714	
Diesel	1.4571	
Liquefied petroleum gas	1.7143	
Refinery dry gas	1.5714	
Blast furnace gas	0.1286	Kilograms of standard coal/ Cubic meter
Coke oven gas	0.5714	
Converter gas	0.2714	
Thermal energy	0.03412	Kilograms of standard coal/Million joules
Electricity	0.1229	Kilograms of standard coal/ Kilowatt-hour

To achieve this, it is essential to first convert the consumption of electricity, thermal energy, and various fossil fuels into a unit measured in standard coal, based on the conversion reference coefficients for standard coal shown in Table 2, and then convert these to terawatt-hours. The specific unit conversion formula is as follows:

$$E_i = E_i' \times \frac{\propto}{\beta} \tag{1}$$

where Ei represents the energy consumption after unit conversion, Ei' represents the energy consumption before unit conversion, α is the standard coal conversion reference coefficient for that energy type, and $\beta=0.1229$Mtce/ TWh.

Table 3 presents some of the cost data used in EnergyPLAN for this study. With continuous technological advancements, the unit investment cost of clean energy is expected to decrease over time.

Table 3. Energy investment and fixed operation and maintenance costs for clean energy generation technologies in 2030 and 2050.

Energy types	Units	Investment cost (MCNY/unit)	Lifespan (years)	Operation and maintenance costs
Onshore wind power	MW_e	0.90	30	2.88
Offshore wind power	MW_e	2.12	40	3.22
Photovoltaic power	MW_e	0.69	30	1.00
Tidal energy	MW_e	5.33	20	3.66
Run-of-river hydropower	MW_e	3.30	50	2.00
Pumped storage power stations	MW_e	7.50	50	1.50
Nuclear power	MW_e	3.02	30	1.96
Interconnected systems	MW_e	1.20	40	1.00
Waste-to-energy combined heat and power	TWh/ year	215.6	20	7.40
Large-scale combined heat and power units	MW_e	0.79	25	3.80
Bioethanol power stations	MW_e	0.44	20	7.70
Electrolytic cells	MW_e	0.28	15	3.00

3. Results

To verify the accuracy and reliability of EnergyPLAN in simulating the operation of Guangxi's energy system, we collected officially released authoritative data to establish a reference model. Figure 5 displays the simulation results of power generation and major primary energy consumption for various power generation technologies in the Guangxi region for 2020. Figure 3 (a) illustrates, through a scatter plot, that the difference between the simulated and actual values is negligible, indicating a high degree of accuracy in the simulation results. Figure 3 (b) shows that most simulated values exhibit an error rate of less than 0.5% compared to actual values, with two samples — external system input electricity and natural gas consumption — showing errors of around ±1.5%. However, the volumes of these two items account for a very small portion of their respective categories, so these errors have a minimal effect on the overall model.

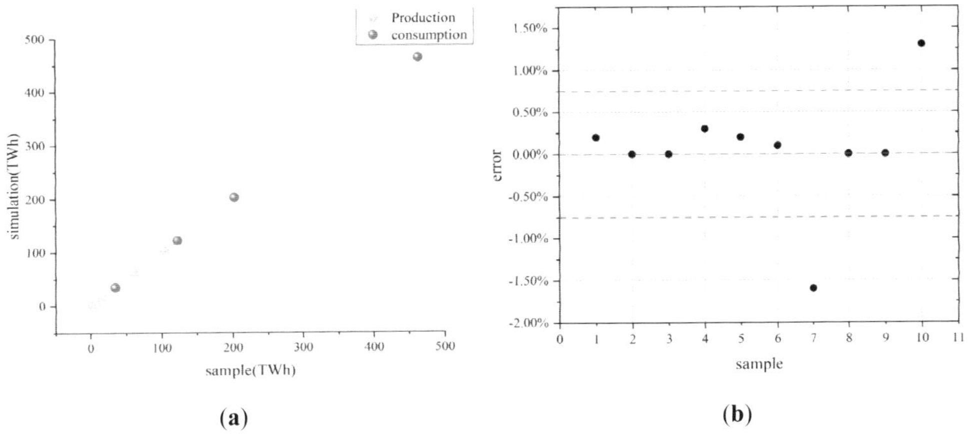

Figure 3. **(a)** Comparison of official data and simulation results; **(b)** Sample error rates.

Journal of Physics: Conference Series **3031** (2025) 012017 doi:10.1088/1742-6596/3031/1/012017

Based on this validation, a quantitative future energy system transition scenario (2050HRES) for the Guangxi region has been established for model optimization using EnergyPLAN software. Table 4 details the specific inputs and technology development plans for 2050, while Table 5 provides the expected consumption of each energy source in the transportation and industrial sectors projected for that year.

Table 4. Planned installed capacity of power generation technologies in Guangxi by 2050.

Power supply type	Installed capacity (MW)
Offshore wind	22500
Onshore wind	70,000
Coal power	25,000
Natural gas-fired power generation	10000
biomass	15000
Run-of-the-river hydroelectricity	27000
Pumped storage	24000
Photovoltaic (e.g. cell)	90000
Nuclear energy	26000
Centralized energy storage	10000MW/20GWh
Electricity consumption of society as a whole	569TWh

Table 5. Energy consumption from different sources in the transportation and industrial sectors in Guangxi by 2050.

Type of energy	Transportation sector (TWh)	Type of energy	Industrial sector (TWh)
Petroleum	6.07	Coals	63.78
Liquefied Petroleum Gas	4.04	Petrochemical	34.02
Diesel	62.68	Petroleum	51.03
Aviation fuel	16.18	Biomass	34.02
Diesel fuel	64.70	Hydrogen	21.26
Electricity	14.15	Electrical power	221.12
Biomass	24.26		
Hydrogen	8.09		

To verify the practicality of this study, we introduce a model, 2050HRES, which has not been optimized by the algorithm, and compare its results with 2050OPT, optimized by the GWO. Figure 5 shows a comparison of cost composition. Table 6 presents a series of key indicators in the 2050OPT scenario compared to the 2050HRES scenario. The optimized 2050OPT scenario shows a 1.8% increase in CO_2 emissions compared to the 2050HRES scenario and an 18.5% decrease significant reduction in the total annual cost of the energy system. This trend results from the fixed costs in the 2050OPT scenario remaining similar to those in the 2050HRES scenario, while variable costs and annual investment costs are substantially lower. This outcome aligns with the goal of minimizing the total annual cost of the energy system using the GWO.

Table 6. Comparison of key indicators in the 2050OPT and 2050HRES scenarios.

Key indicators	2050OPT	2050HRES
Carbon dioxide emissions (millions of tons)	95.4	93.7
Total annual cost (millions of dollars)	41713	51198
Share of RES in primary energy (%)	50.5	50.6
Share of RES in electricity production (%)	84.2	85.0
Electricity production by RES (TWh)	513.7	515.0
Share of clean energy in primary energy (%)	69.5	69.7

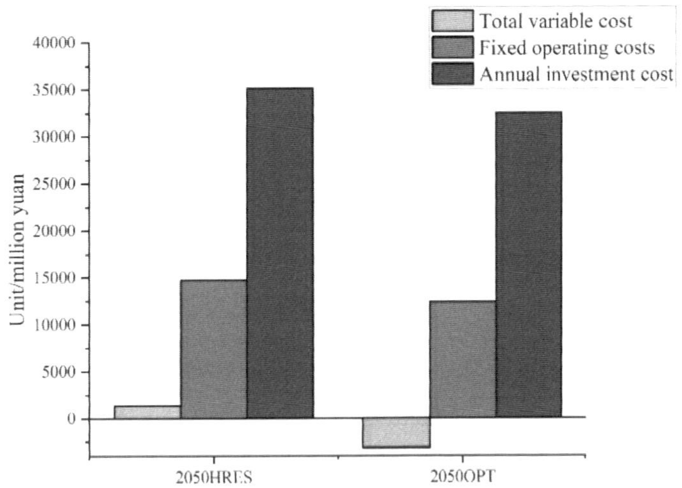

Figure 4. Energy system cost composition in 2050HRES and 2050OPT.

In the 2050OPT scenario, the configuration of installed capacity closely resembles that of the 2050HRES scenario, with fossil power generation accounting for about 11% of the total. Meanwhile, renewable and clean energy sources are projected to reach approximately 80% and 89% of the installed capacity, as illustrated in Figure 4. This optimized configuration of the 2050OPT scenario not only demonstrates the possibility of transitioning to a low-carbon and high-efficiency energy system but also reflects the ability to optimize the GWO to achieve a balance between economic and environmental benefits.

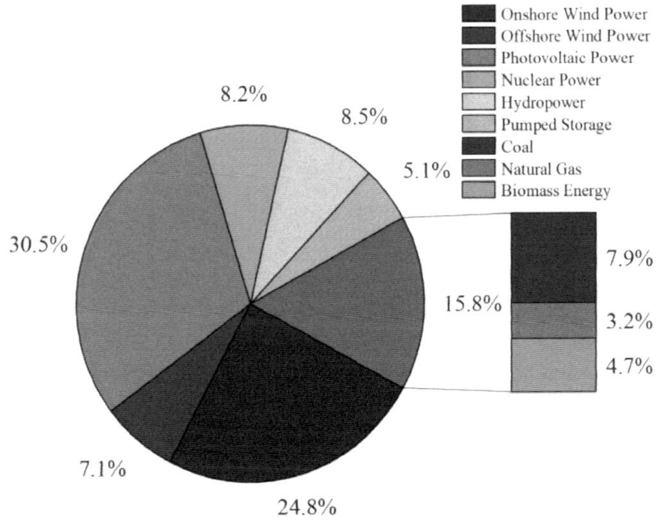

Figure 5. Installed capacity of Guangxi electric power system in 2050OPT scenario.

4. Conclusion

The results demonstrate that the optimization scenario effectively achieves an 18.5% reduction in total system costs while maintaining a comparable level of CO_2 emission reduction, thus fulfilling the goal of optimal economic cost-effectiveness in future energy system configuration. Furthermore, clean energy is projected to dominate the energy structure of the Guangxi region, expected to satisfy about 84.2% of the region's electricity demand.

Journal of Physics: Conference Series **3031** (2025) 012017 doi:10.1088/1742-6596/3031/1/012017

References

[1] Y. Li, T. Liu, J. Xu, Analyzing the economic, social, and technological determinants of renewable and nonrenewable electricity production in China: Findings from time series models, Energy 282 (2023).
[2] L.S. Plaga, V. Bertsch, Methods for assessing climate uncertainty in energy system models systematic literature review, Applied Energy 331 (2023).
[3] M.G. Prina, B. Nastasi, D. Groppi, S. Misconel, D.A. Garcia, W. Sparber, Comparison methods of energy system frameworks, models and scenario results, Renewable & Sustainable Energy Reviews 167 (2022).
[4] H. Lund, J.Z. Thellufsen, P.A. Ostergaard, P. Sorknaes, I.R. Skov, B.V. Mathiesen, EnergyPLAN e Advanced analysis of smart energy systems, Smart Energy 1 (2021).
[5] M.G. Prina, M. Lionetti, G. Manzolini, W. Sparber, D. Moser, Transition pathways optimization methodology through EnergyPLAN software for long-term energy planning, Applied Energy 235 (2019) 356-368.
[6] S. Ferrari, F. Zagarella, P. Caputo, M. Bonomolo, Assessment of tools for urban energy planning, Energy 176 (2019) 544-551.
[7] O. Bamisile, X. Wang, H. Adun, C.J. Ejiyi, S. Obiora, Q. Huang, W. Hu, A 2030 and 2050 feasible/sustainable decarbonization perusal for China's Sichuan Province: A deep carbon neutrality analysis and EnergyPLAN, Energy Conversion and Management 261 (2022).
[8] M. Yuan, J.Z. Thellufsen, P. Sorknaes, H. Lund, Y. Liang, District heating in 100% renewable energy systems: Combining industrial excess heat and heat pumps, Energy Conversion and Management 244 (2021).
[9] J. de Maigret, D. Viesi, M.S. Mahbub, M. Testi, M. Cuonzo, J.Z. Thellufsen, P.A. Ostergaard, H. Lund, M. Baratieri, L. Crema, A multi-objective optimization approach in defining the decarbonization strategy of a refinery, Smart Energy 6 (2022).
[10] S. Dereli, A new modified grey wolf optimization algorithm proposal for a fundamental engineering problem in robotics, Neural Computing & Applications 33(21) (2021) 14119-14131.
[11] H. Faris, I. Aljarah, M.A. Al-Betar, S. Mirjalili, Grey wolf optimizer: a review of recent variants and applications, Neural Computing & Applications 30(2) (2018) 413-435.
[12] S. Mirjalili, S.M. Mirjalili, A. Hatamlou, Multi-Verse Optimizer: a nature-inspired algorithm for global optimization, Neural Computing & Applications 27(2) (2016) 495-513.

AUTHOR INDEX